Engineering Materials

This series provides topical information on innovative, structural and functional materials and composites with applications in optical, electrical, mechanical, civil, aeronautical, medical, bio- and nano-engineering. The individual volumes are complete, comprehensive monographs covering the structure, properties, manufacturing process and applications of these materials. This multidisciplinary series is devoted to professionals, students and all those interested in the latest developments in the Materials Science field, that look for a carefully selected collection of high quality review articles on their respective field of expertise.

Indexed at Compendex (2021) and Scopus (2022)

Damian Batory

Tribology of Low Friction Carbon Based Coatings

Springer

Damian Batory
Department of Vehicles and Fundamentals
of Machine Design
Lodz University of Technology
Lodz, Poland

ISSN 1612-1317   ISSN 1868-1212 (electronic)
Engineering Materials
ISBN 978-3-031-95978-3   ISBN 978-3-031-95979-0 (eBook)
https://doi.org/10.1007/978-3-031-95979-0

© The Editor(s) (if applicable) and The Author(s), under exclusive license to Springer Nature Switzerland AG 2025

This work is subject to copyright. All rights are solely and exclusively licensed by the Publisher, whether the whole or part of the material is concerned, specifically the rights of translation, reprinting, reuse of illustrations, recitation, broadcasting, reproduction on microfilms or in any other physical way, and transmission or information storage and retrieval, electronic adaptation, computer software, or by similar or dissimilar methodology now known or hereafter developed.
The use of general descriptive names, registered names, trademarks, service marks, etc. in this publication does not imply, even in the absence of a specific statement, that such names are exempt from the relevant protective laws and regulations and therefore free for general use.
The publisher, the authors and the editors are safe to assume that the advice and information in this book are believed to be true and accurate at the date of publication. Neither the publisher nor the authors or the editors give a warranty, expressed or implied, with respect to the material contained herein or for any errors or omissions that may have been made. The publisher remains neutral with regard to jurisdictional claims in published maps and institutional affiliations.

This Springer imprint is published by the registered company Springer Nature Switzerland AG
The registered company address is: Gewerbestrasse 11, 6330 Cham, Switzerland

If disposing of this product, please recycle the paper.

Competing Interests The author has no competing interests to declare that are relevant to the content of this manuscript.

Introduction

Since the dawn of time, friction has accompanied humanity in every aspect of technical and daily life. At first glance, the seemingly simple and often unnoticed abilities to walk or grasp various objects are made possible, in part, by the phenomenon of friction. Friction is intrinsically connected to other physical phenomena, such as heat generation, vibrations, and wear. Over the centuries, people have learned to harness friction for their own purposes, from lighting fires by rubbing sticks together to producing sound from stringed instruments.

Friction plays a crucial role in the operation of mechanical systems, particularly in those where components slide or roll against one another. This results in motion resistance at the boundaries of interacting surfaces or from the environment in which a given object moves. Friction is fundamental to the functioning of vehicle drive and braking systems: tire friction on the ground provides drivers with control, allowing for cornering and stopping. Advanced material engineering technologies also make use of the heat generated by friction between two elements in welding processes. Low friction resistance, on the other hand, is essential for the smooth operation of joints, various moving mechanisms, bearings, and gears.

The mechanical contact between interacting surfaces generates drag force, which is the primary cause of wear processes involving the progressive removal of material. In some engineering applications, however, controlled wear is desirable and, in certain cases, it is the primary goal of the operation. Among the many technologies that utilize wear processes are manufacturing methods such as milling, cutting, and electro-erosion, as well as finishing techniques like polishing. Generally, however, wear is an undesirable phenomenon that accompanies the operation of machines and devices.

A range of mechanisms drive progressive wear processes, and numerous methods are available to reduce or even eliminate them. These methods include the use of effective lubrication systems and surface modifications on friction contact elements to decrease frictional force and surface wear, thereby extending service life and enhancing the efficiency of the entire system.

The field of science and technology that focuses on interactions between cooperating surfaces and related phenomena is known as tribology. Since friction and wear are often sources of significant economic, design, and environmental challenges, tribology is a crucial tool for driving continuous technological advancement. Fundamentally, it represents a synergy of knowledge, skills, and experience that integrates disciplines such as mechanical engineering, materials science, physics, chemistry, and medicine in an interdisciplinary approach.

Today, we are confronted with new challenges, such as sustainable development, climate change, and the gradual degradation of the environment. These issues demand innovative solutions and focused thinking. As humanity advances, tribology plays a crucial role in addressing the demand for cutting-edge technologies, contributing to outcomes such as reduced fuel consumption, lower greenhouse gas emissions, enhanced durability of machinery and devices, and improved quality of life through advancements like medical implants.

Surface modification and the refinement of machine parts, engines, and biomaterials rank among the major achievements of materials engineering in the past century. A wide range of surface engineering technologies, at relatively low costs, enables significant extension of service life and enhancement of mechanical, tribological, and corrosion-resistant properties of machine components, devices, and medical implants operating under high loads, friction, or in harsh environments. Current research on surface modification is centered on designing and producing multifunctional materials and coatings with exceptional performance characteristics. These coatings often combine high hardness with other desirable properties, such as a low coefficient of friction and high resistance to wear and corrosion, maintained across both low and high operating temperatures.

Fairly extensive literature on tribology of metal materials, polymers, ceramics and functional coatings focus on general aspects of their friction and wear, including contact mechanics, as well as synthesis and application possibilities. It undoubtedly provides a rich source of information on methods and techniques for providing low friction resistance and the highest possible protection against wear.

The objective of the author of this book is to present a comprehensive characterisation of low-friction layers, with a particular focus on diamond-like coatings. In addition to an overview of the frictional properties of carbon coatings, the reader will also find a review of the most commonly used methods for testing their tribological properties. Ultimately, a range of tools and test methods for the surface analysis of the obtained wear scars and wear tracks will be presented. The book could be a valuable resource for Ph.D. students and young scientists starting their scientific adventure with low-friction carbon based materials. By providing detailed insights into coating technologies, testing techniques, and analytical tools, such a guide can help them develop a strong foundation in surface engineering and tribology. It will empower researchers to select the right coatings and testing methods for their specific applications. In addition, it will enable them to conduct experiments with greater accuracy and efficiency, so that research conducted and the scientific papers prepared on their basis shall be of the highest scientific standard.

Contents

1 General Characteristics of Friction and Wear of Engineering Materials ... 1
  1.1 Friction Contact and Its Tribological Characteristics ... 1
  1.2 Sliding and Rolling Friction ... 4
  1.3 Resistance to Motion in Sliding Friction ... 6
  1.4 Basic Friction and Wear Mechanisms of Coatings ... 12
  1.5 Coatings Tribology at Scale ... 21
    1.5.1 Nanoscale Mechanisms ... 25
    1.5.2 Microscale Mechanisms ... 27
    1.5.3 Macroscale Mechanisms ... 30
  References ... 32

2 Solid Lubrication ... 37
  2.1 Solid Lubricant Coatings ... 40
    2.1.1 Polymer Coatings ... 43
    2.1.2 Soft Metal Coatings ... 44
    2.1.3 Lamellar Solids ... 45
    2.1.4 Self-healing Coatings ... 47
    2.1.5 Self-organized, Adaptive Chameleon Coatings ... 49
  2.2 Synergy of Solid Lubricants and Surface Texturing ... 53
  References ... 58

3 DLC (Diamond-Like Carbon Coatings) ... 63
  References ... 68

4 Tribological Properties of DLC Coatings ... 71
  4.1 The Influence of the Surface Geometrical Structure of DLC Coatings on Their Tribological Properties ... 72
  4.2 Saturation of Dangling Bonds of Carbon Atoms at the Sliding Interface with Passivating Species (Passivation Mechanism) in the Tribology of Diamond-Like Carbon Coatings ... 75

	4.3	The Role of Stress and Shear Induced (Tribo-Induced) Graphitization in Tribology of Diamond-Like Carbon Coatings	79

- 4.3 The Role of Stress and Shear Induced (Tribo-Induced) Graphitization in Tribology of Diamond-Like Carbon Coatings 79
- 4.4 The Role of Third-Body Interactions in Tribology of Diamond-Like Carbon Coatings 83
- References 89

5 Doped Diamond-Like Carbon Coatings with Metallic and Nonmetallic Elements 93
- 5.1 Fluorinated DLC 94
- 5.2 Silicon Incorporated Carbon Coatings 96
- 5.3 Metal Incorporated DLC 99
- References 104

6 Methods of Testing Friction and Wear of Coatings 109
- 6.1 Tribological Investigations 111
- 6.2 Tribotesters 113
  - 6.2.1 A Device with a Pin/Ball-on-Disc Configuration 114
  - 6.2.2 A Device with Block-on-Ring Configuration 116
  - 6.2.3 Nanoscale Tribology 118
  - 6.2.4 Fatigue Analysis by Micro and Nanoimpact Testing 119
  - 6.2.5 Microabrasion Wear Testing (Calotest) 121
  - 6.2.6 A Device with Pin and Vee-Block Configuration 123
  - 6.2.7 A Device with Three Rolls and Cone Configuration 124
  - 6.2.8 Three-Ball-on-Rod Rolling Contact Fatigue (RCF) Testing Machine 125
  - 6.2.9 Micropitting Studies of Coated Surfaces in Rolling/Sliding Contacts 127
  - 6.2.10 A Device with a Pair of Spur Gears 131
  - 6.2.11 Other Custom Made Tribometers 135
- References 138

7 Methods of Characterization of Carbon Based Low Friction Coatings After Tribological Testing 143
- 7.1 Profilometry 143
- 7.2 Optical and Scanning Electron Microscopy 145
- 7.3 X-Ray Photoelectron Spectroscopy 147
- 7.4 Raman Spectroscopy 149
- 7.5 Fourier Transform Infrared Spectroscopy 153
- 7.6 In-Situ Studies of Tribological Processes of Carbon Based Low Friction Coatings 154
- References 157

8	**Tribocorrosion and Biotribocorrosion of Carbon Based Low Friction Coatings**	161
	References	167
9	**Synergistic Effect of Carbon Based Coatings with Self-assembled Monolayers**	169
	References	175

Chapter 1
General Characteristics of Friction and Wear of Engineering Materials

Abstract This chapter provides a comprehensive overview of the subject of friction and wear of engineering materials. In addition to other topics, the chapter presents a detailed discussion of the characteristics of friction contact. It also discusses the origins of resistance to sliding friction and the methods of reducing this resistance. The chapter then goes on to explore the mechanisms of wear of coating materials. Finally, it presents different approaches in terms of the scale of identification and understanding the tribological processes of coating materials.

1.1 Friction Contact and Its Tribological Characteristics

The movement of one solid surface relative to another is fundamentally important for the function of many mechanisms, both artificial and natural. Mechanical systems where the surfaces of cooperating elements remain stationary relative to each other are rare. In most cases, cooperating elements form a frictional contact, whose characteristics are mainly determined by the intended purpose and are defined precisely by the individual components of a given mechanism. Frictional contact occurs when two nominally flat and parallel surfaces come into contact. At the microscale, this contact is established in micro-areas via the asperity peaks of the rough surfaces. Since these uneven areas are the only points of contact between the cooperating surfaces, they are responsible for load transfer and also generate frictional forces. The area or number of contacting asperities forming the frictional contact is a function of the applied load and depends directly on the mechanical properties of the materials involved. Therefore, understanding the interactions between contacting asperity peaks under applied forces is fundamental to any study of friction or wear. It can also be assumed that surfaces forming a frictional contact have high smoothness. In such cases, the macroscopic distribution of elastic stresses near the contact area provides valuable insights into wear mechanisms like plastic deformation, cracking, or fatigue [1–3]. Both types of frictional contacts are reflected in real-world mechanical device solutions. The intended function, purpose, and kinematic characteristics of the contact allow friction nodes to be categorized as ***conformal*** or ***non-conformal***. The first

© The Author(s), under exclusive license to Springer Nature Switzerland AG 2025
D. Batory, *Tribology of Low Friction Carbon Based Coatings*, Engineering Materials, https://doi.org/10.1007/978-3-031-95979-0_1

group includes friction nodes with a large nominal contact area, such as loosely or slidingly fitted connections, systems used in brakes or sliding bearings, as well as in hip joint implants or spinal intervertebral disc prostheses. In these nodes, due to the large and generally evenly distributed contact surface, the primary source of friction forces is the interaction of roughness peaks. Contact pressures in this type of friction node are relatively low, with the risk of exceeding the plastic limit occurring only in the micro-areas where surface asperities interact [2–4]. Non-conformal friction contact, also known as concentrated contact, occurs in bearings, rolling guides, cam mechanisms, gears, and precision devices operating under drilling friction conditions. The contact between the cooperating surfaces is point or linear. In reality, however, due to elastic deformation of the material near the contact area, it becomes two-dimensional with a very small nominal area. As a result, even a slight load generates rolling friction resistance, and the high concentration of stress associated with this contact often leads to surface damage in the cooperating elements, most commonly of a fatigue nature. The analysis of elastic stress fields in concentrated contact was first explored by Heinrich Hertz in 1881, and these stresses are now referred to as Hertzian stresses. A more detailed discussion of the theory of contact in friction pairs, along with their specific characteristics and methods for analysis and modeling, can be found in the literature [1, 3, 5, 6].

Up to this point, the discussion regarding the contact of cooperating surfaces has focused on homogeneous materials with a relatively high modulus of elasticity. However, many components of machines and devices undergo surface treatments designed to modify their functional properties. Increasingly, various hard technical coatings resistant to wear from friction and other forms of surface damage are employed, along with low-friction coating materials aimed at reducing the coefficient of friction in frictional contacts. These advanced surface engineering technologies allow for the modification of treated surface properties with high accuracy and predictability, achieving molecular or even atomic precision. Figure 1.1 illustrates the spectrum of possibilities for shaping the properties of the substrate material through the surface modification techniques, whereas Fig. 1.2 shows the schematic of surface modified material under friction conditions. The selection of technology, process parameters, coating material substrates, and activation methods creates extensive opportunities for establishing surface tribological properties related to friction and wear in various industrial applications.

Despite the ability to define the required values for the friction coefficient or wear resistance for a given application relatively simply, the challenge lies in the lack of effective general tools for predicting the basic properties of coatings. Key parameters include thickness, surface geometrical structure, material composition and hardness, Young's modulus, residual stresses, and crack resistance. The substrate material is also critical, as is the need to use an intermediate coating that enhances adhesion. The solutions for the contact stresses in friction pairs—where one or both materials are modified by coatings—are complex and often require numerical methods such as the Finite Element Method (FEM). However, if the coating thickness is significantly smaller than the depth at which shear stresses are the highest (dependent on contact geometry and loading conditions), it may be reasonable to exclude the coating from

1.1 Friction Contact and Its Tribological Characteristics

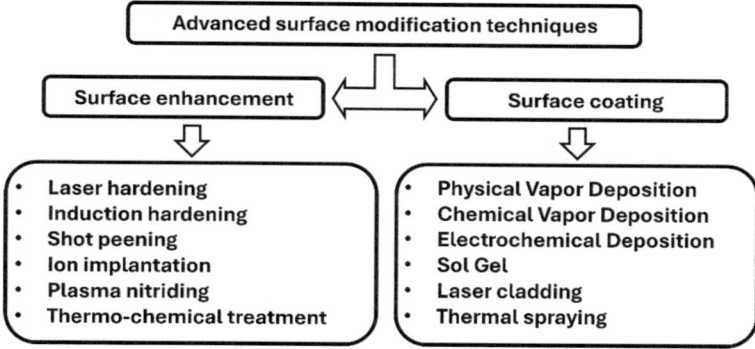

Fig. 1.1 Advanced surface treatment techniques offering wide possibilities for modifying and adjusting mechanical and chemical properties of the material for a specific application

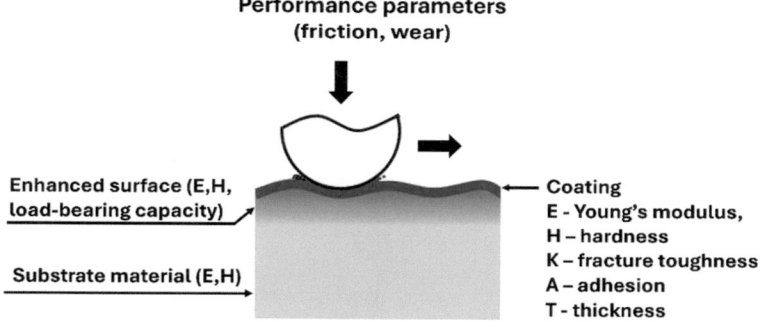

Fig. 1.2 Schematic illustration of the structure of a surface-modified material designed for tribological application

the analysis and instead apply Hertz's formulas along with the material data of the cooperating substrates [1]. Nevertheless, as noted in other literature, Hertzian theory can be adapted to the case of concentrated contact in coated elements by integrating it with a numerical model [7–9].

Hard technical coatings are characterized by high brittleness, making them susceptible to crack formation and propagation. In sliding contacts, high compressive stress at the contact edge can lead to brittle damage in the coating. Cracks may develop in coatings due to elastic and plastic deformation of the substrate material, particularly when the substrate's mechanical properties are insufficient to provide a solid foundation for the hard and brittle coating. Another form of coating failure is delamination, which manifests as splintered fragments of the coating. When a normal force loads the friction node, the area of highest contact stress concentration is located at a depth significantly greater than the range of the substrate-coating interface. However, under tangential loads, typical of sliding contact, this depth decreases, and the point of maximum contact stress shifts closer to the surface. For instance, when the ratio of tangential to normal load is 0.25, this point will be at the surface [1].

1.2 Sliding and Rolling Friction

The friction force can be defined as the resistance encountered by one body when moving across the surface of another. This definition encompasses two cases of relative motion: the first is **sliding friction**, and the second is **rolling friction**. Sliding friction primarily occurs in congruent friction pairs, meaning in conformal contact, while rolling friction is characteristic of non-conformal friction pairs. However, these two phenomena are not mutually exclusive; machine elements in concentrated contact often operate under rolling friction conditions while also experiencing elements of slip or sliding friction. In both cases—ideal sliding friction and ideal rolling friction—a tangential force (F) is required to move one body relative to the stationary counterpart, which is equal in magnitude to the oppositely directed friction force (F_t) (see Fig. 1.3a). Therefore, the coefficient of sliding friction can be defined as the ratio of the friction force (F_t) to the normal force (N) exerted by the body on the surface over which it moves:

$$\mu = \frac{F_t}{N} \tag{1.1}$$

In the case of rolling friction (Fig. 1.3b), the vertical component of the reaction force (R) balances the normal force (N), while the horizontal component (F_t) counteracts the pulling force (F), allowing the body to remain at rest or move with uniform motion. From the equilibrium of moments around the center of the cylinder, the following relationship can be established:

$$F_t \cdot r = R \cdot f \tag{1.2}$$

and after transformation:

$$f = \frac{F_t \cdot r}{R} \tag{1.3}$$

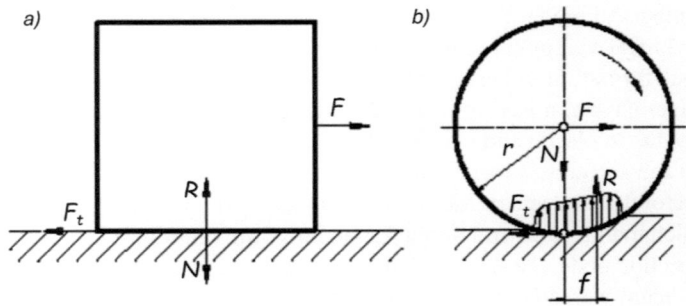

Fig. 1.3 Graphical interpretation of **a** the sliding and **b** rolling friction coefficient

1.2 Sliding and Rolling Friction

Similarly to sliding friction, the rolling friction coefficient is defined as the ratio of the friction force to the normal force.

The magnitude of the friction force is determined by the coefficient of friction, which can vary widely—from 0.001 for lightly loaded rolling bearings to over 10 for atomically pure metal surfaces interacting under vacuum conditions. In general, the coefficient of rolling friction is one to two orders of magnitude lower than that of sliding friction, ranging from 0.005 to 0.05. Therefore, in principle, there is no technical problem in lowering the resistance of rolling friction, only the need to ensure its realisation and the durability of the friction node.

In the case of sliding friction, which is particularly relevant for the application of low-friction coating materials, the two main causes of friction resistance are the adhesion forces at the actual contact areas of the asperity peaks and the additional forces generated during the attempt to move one of the friction pair elements tangentially. In this process, the hard asperity peaks, pressed into the soft material of the counterbody by the normal force, cause micro-cutting or ploughing. On a microscopic scale, friction is a stochastic process composed of various phenomena. The occurrence and intensity of these phenomena at different locations on the friction contact surface depend on random variables, such as the distribution of surface asperities, which is influenced by the surface geometrical structure [1, 3, 4, 10].

Friction has been studied for centuries. The first to research this phenomenon was Leonardo Da Vinci, who observed the proportionality of the friction force to the normal force. Two centuries later, Guillaume Amontons conducted research on friction. He reached similar conclusions to his predecessor, confirmed over the following years. In this way, two laws of friction were formulated:

- *The friction force is directly proportional to the normal force.*
- *The friction force does not depend on the area of contact of the bodies.*

The first one can therefore be written in the following form:

$$F_t = \mu \cdot N \qquad (1.4)$$

which is also a statement that the coefficient of friction is independent of the applied load and is the commonly known Amontons expression. For many materials, both under dry friction conditions and for lubricated friction contacts, it is a fairly faithful approximation of phenomena occurring in macroscopic contact areas. The second law, although not as widely studied and developed as the first, has nevertheless been confirmed for most engineering materials with the exception of polymers.

The common feature of the studies presented above was the assumption that the only source of friction is the unevenness of the mating surfaces. This would mean that the friction force can be practically eliminated by smoothing the contact surfaces. This was in contradiction with experiments where two metal surfaces were polished, thereby increasing their friction force during dry sliding against each other.

The next scientist to work on the phenomena of friction was Coulomb. His work proved that the friction force has two components: the first dependent on the load and the second, related to the adhesion forces of two surfaces in frictional contact. This research led to the formulation of the third law of friction:

- *The friction force is independent of the sliding speed*

This law was originally based on the measurement of the force required to initiate movement, which usually has a higher value than that required to maintain it, and states that the coefficient of static friction has a higher value than the coefficient of dynamic (kinetic) friction. However, when a friction node is put into motion, the coefficient of dynamic friction for many systems is almost independent of the sliding speed over a wide range. Since the mechanical energy in friction contacts operating under dry friction conditions is dissipated in the form of heat, there are exceptions to this rule, primarily for tribological systems operating under high relative speed conditions.

Over time, the scientific community has come to conclusion that the influence of adhesion forces should be taken into account. The currently prevailing models have become standardized, stating that friction can be called a set of phenomena occurring in the area of contact between two bodies moving relative to each other, as a result of which resistance to motion is created [10].

1.3 Resistance to Motion in Sliding Friction

Let us return to the discussion of the sources of resistance to motion. Both the works of Amontons and Coulomb assumed at their foundation that the main source of friction forces are mechanical interactions between rigid and elastically deformed surface asperities. Figure 1.3 shows a simplified diagram of this model assuming the climbing of one sawtooth surface onto another. It can be easily seen, by comparing the work done by the friction force (T) to the work done against the normal force (N), that the coefficient of friction μ is equal to the tangent of the angle Θ (Fig. 1.4a). However, at the next stage it is equally easy to notice the basic flaw of this model. After the upper body reaches the highest point (Fig. 1.4b), in which we observe direct contact of the vertices of the surface irregularities, the normal force causes the upper body to start sliding down (Fig. 1.4c), and therefore work is done, and the potential energy accumulated in the first phase is recovered. According to the presented model, in the entire cycle of movement of the upper body relative to the lower one, there are no energy dissipation processes, and therefore on a macroscopic scale there should be no friction force either.

Undoubtedly, the existence of some energy dissipation mechanism is a necessary condition for any satisfactory friction model. Nowadays, the geometrical structure of the surface (so-called roughness theory) is no longer considered as the main source of friction in technical applications.

Among the series of theoretical friction models, the dry friction model proposed by Bowden and Tabor [4] is used in modern technology. This model in its simplest form was developed for metals and even in their case it has limited applicability. It

1.3 Resistance to Motion in Sliding Friction

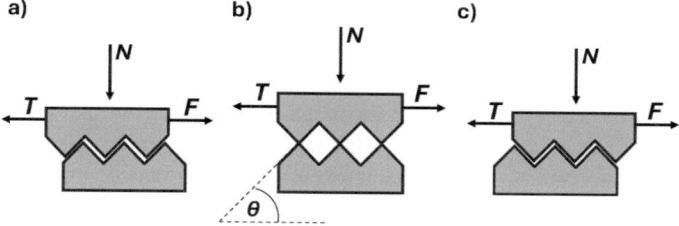

Fig. 1.4 Schematic illustrating the interaction of two surfaces with sawtooth geometry

assumes that the friction force is a resultant of the following components: adhesive force (T_{adh}) generated in the areas of actual contact of the surface asperities and plastic deformation (T_{pl}) caused by the roughness peaks of the harder material, which are pressed by the normal force into the surface of softer material. Under the influence of the applied tangential forces these surface asperities may also cause groove formation in the surface of softer material:

$$T = T_{adh} + T_{pl} \tag{1.5}$$

In later developments of this theory, it became obvious, that the two components cannot be perceived and considered as independent.

The adhesive component results from intermolecular interactions, which are assumed to arise in the immediate vicinity of the roughness peaks of the contacting metals. However, many experiments in this area have shown that the value of the adhesion force can vary within wide limits. As previously mentioned, assuming that the main component of the friction force is the surface geometrical structure of the friction contact elements, it would be expected that two polished surfaces should be characterized by a very low friction coefficient. However, practice has shown that the friction coefficient of such a pair was significantly higher compared to the initial variant. On the other hand, under normal conditions, two metal surfaces loaded with a normal force generally do not exhibit noticeable adhesion forces. The difference between these two cases is twofold. The first to be mentioned is the chemical structure of the contacting surfaces. In reality, surfaces are covered with layers of oxides and other adsorbed functional groups originating from the air or atmosphere prevailing in the working conditions of a given element. This significantly reduces the forces of mutual intermolecular interactions, thereby reducing adhesion forces. The second component is elastic deformation around the asperity peaks, which, under the influence of load, generate stresses with values significantly exceeding the cohesion forces of the resulting adhesive bonds. The fact is, however, that if two different metal surfaces, at least partially free of surface oxide layers and adsorbed gases and impurities, are brought close to the distance of the interaction of intermolecular forces, the contacting asperity peaks will create very strong adhesive bonds. Additionally, in the case of mutual movement of these surfaces, friction microwelds may be crated. The bond energy in these connections significantly exceeds the shear strength of the softer material. This leads to the tearing out of its fragments and their transfer to the harder surface of the cooperating tribo-pair. These places are particularly susceptible to the formation of new adhesive bonds, and the progressive processes of destruction and transfer of material in the friction pair lead to accelerated wear of the surface of lower hardness. Intensification of this process on a macroscopic scale can lead to the most pathological form of wear, i.e. seizure, and thus permanent immobilization of the tribological system [2, 3, 11].

If we assume that the actual contact area is the sum of the cross-sectional areas of all adhesive joints (A) and that all joints are characterized by the same value of shear strength (R_s), the adhesive component of the friction force can be written as:

$$T_{adh} = AR_s \tag{1.6}$$

It is also known, that in contact conditions, depending on the load on the friction pair, the roughness peaks of the cooperating materials can deform both elastically and plastically. The actual contact surface shows an almost linear relationship with the applied normal force. Let us consider two metal surfaces of different hardness machined with commonly used engineering methods. As a result of loading the friction pair with a normal force (N), the initial nature of the contact between the cooperating surfaces will bear the characteristics of plastic deformation. Therefore, the value of normal stresses transferred by the roughness peaks will be close to the hardness of the softer material (H), and the approximate area of the actual contact surface will be:

$$A \approx \frac{N}{H} \tag{1.7}$$

The adhesion component of the friction coefficient resulting from the adhesion forces is equal to:

$$\mu_{adh} = \frac{T_{adh}}{N} \tag{1.8}$$

And after substitution to Eq. 1.6, we obtain:

$$\mu_{adh} \approx \frac{R_s}{H} \tag{1.9}$$

The newly created frictional connections are torn apart in the volume of the weaker material. It can therefore be assumed with great approximation that R_s is equal to the shear strength of this material. In the case of metals, hardness is about three times the yield strength [12, 13], therefore we can write that:

$$H \approx 3R_e \tag{1.10}$$

1.3 Resistance to Motion in Sliding Friction

For isotropic materials the yield strength is approximately 1.7 to 2 times the shear strength, so:

$$H \approx 5R_s \tag{1.11}$$

substituting back into Eq. 1.6, we obtain:

$$\mu_{adh} \approx \frac{R_s}{H} \approx 0.2 \tag{1.12}$$

The component of plastic deformation related to the micro-cutting or ploughing of the soft material in the friction pair by the hard asperity peaks of the counterbody surface can be determined using a simple model of a single roughness element with an idealized shape. Let us consider the diagram presented in Fig. 1.5.

A rigid cone with a half-opening angle φ moves across a flat surface of plastically deformable material. To move the indenter and achieve plastic deformation, it is necessary to apply a tangential force that will generate stresses exceeding the yield strength, specifically equal to the hardness of the deformable substrate ($H = 3R_e$). The friction force for such a surface is equal to the area of the resulting groove multiplied by the hardness of the soft material:

$$F_{def} = Ha^2 ctg\varphi \tag{1.13}$$

The normal force acting on the cone can be determined from the relationship:

$$N = \frac{H\pi a^2}{2} \tag{1.14}$$

knowing the normal and friction force, we calculate the component of the friction coefficient due to plastic deformation as:

$$\mu_{def} = \frac{F_{def}}{N} = \frac{2}{\pi} ctg\varphi \tag{1.15}$$

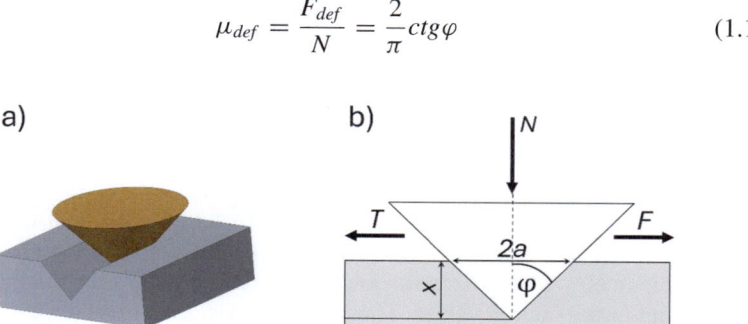

Fig. 1.5 Process of plastic deformation in a soft material during sliding motion: **a** isometric view; **b** cross-section diagram [1, 10]

The parameters of the geometrical structure of metallic surfaces used in typical engineering applications assume that the angles of inclination of the surface irregularities do not exceed 10°, which in the discussed model translates to values of angle $\varphi > 80°$. Therefore, from expression (1.15), it can be inferred that the contribution of the coefficient of friction due to plastic deformation should not exceed 0.1. In a cumulative view of these two components, one would expect that the coefficient of friction in the case of a hard surface interacting with a soft counter-sample should be approximately 0.3. Furthermore, for a friction pair in which the cooperating surfaces are made of the same material, the component due to plastic deformation becomes marginal. Thus, one could expect that in this case, the coefficient of friction would be even lower (around 0.2). However, both practical observations and numerous literature reports do not support this theory. For a range of metallic friction pairs, both homogeneous and heterogeneous, the recorded values of the coefficient of friction significantly deviate from these expected, sometimes yielding results that far exceed those predicted by the presented model.

It is evident that, in addition to the mentioned adhesive and plastic deformation components, other factors play a significant role in shaping the kinematic characteristics of friction nodes. The two most commonly encountered factors are related to strengthening due to plastic deformation and the actual contact area. Since nearly all metallic materials exhibit a tendency to strengthen due to plastic deformation under applied force, the shear strength of the formed asperity contacts significantly increases relative to the hardness of the softer material. Thus, it can be assumed that strengthening through deformation will increase the adhesive component of the coefficient of friction, however, this phenomenon is difficult to quantify. The final element, therefore, is the real contact area, or more precisely the junction growth.

It is important to note that the aforementioned considerations assumed the contact area to be constant and determined solely by the normal load, without accounting for tangential forces. This is a significant simplification, as according to the yield criteria for elastoplastic materials, the transition to a plastic state depends on both normal and shear stress values. Assuming that the actual contact area can increase with increasing tangential force, it can be concluded that the entire surface of the analysed friction pair would constitute the actual contact area. This would consequently cause the friction coefficient to reach very high values. In the case of certain metals under ultra-high vacuum conditions, this phenomenon can indeed reach such a scale [1]. Nevertheless, in most typical engineering applications, this phenomenon is limited by the mechanical properties of the material, as well as by the formation of thin interfacial layers on the surface, resulting from processes such as oxidation. When the shear strength of the interfacial layer material τ_i is equal to that of the substrate material τ_p, it does not significantly impact the potential to reduce the friction coefficient. However, a decrease in the τ_i/τ_p ratio provides measurable benefits for achieving lower friction values. Thus, if the surface asperities of one element in a friction pair can be isolated from the counterbody by a thin interfacial layer with a shear strength τ_i, then the resulting friction force is the product of the shear strength of the softer material and the actual contact area:

1.3 Resistance to Motion in Sliding Friction

$$F_t = A\tau_i \tag{1.16}$$

The normal load can be expressed as the plastic flow stress of the bulk of asperity (where p_0 is its yield stress in compression):

$$W = Ap_0 \tag{1.17}$$

and thus we can write that:

$$\mu = \frac{F_t}{W} = \frac{\tau_i}{p_0}. \tag{1.18}$$

Expression (1.18) is a key element in considerations regarding the possibilities of reducing sliding friction resistance from both a design and technological perspective. The shear strength and yield strength of the softer material significantly influence the value of the friction coefficient. As previously presented based on the material yield hypotheses, both of these parameters are proportionally interdependent, making it impossible to simultaneously decrease the shear strength while increasing the yield strength. According to the reasoning outlined above, as well as the interpretative model proposed by Bowden and Tabor [4], the use of a counterbody made of a harder material in the friction pair will allow for a slight but measurable reduction in the actual contact area. However, as mentioned earlier, this approach will also lead to an increase in shear strength, which is inherently linked to the energy of chemical bonds in the material and, thus, its hardness. Nevertheless, if we assume that the hard substrate material can serve as a solid foundation for a thin coating of a material characterized by low shear resistance or is surface-neutral concerning its ability to form chemical bonds, the properties of the resulting structural composition will meet all the previously set conditions. Therefore, we obtain a component in the friction contact that, due to the high hardness of the substrate material, is not susceptible to large elastic deformations under the influence of applied normal forces, while at the same time, the low shear strength of the coating material ensures low frictional resistance. A schematic of Bowden and Tabor concept is presented in Fig. 1.6. This solution, in its original form, has found numerous applications in the production of sliding bearings [2, 14]. Further considerations of solid lubrication theory can be found in the following chapters.

The dynamic development of plasma technologies for synthesizing carbon-based thin films, due to their unique mechanical and tribological properties, has led to the creation of another structural composition based on a hardened substrate material, modified with a thin coating of diamond-like carbon that exhibits low chemical affinity to most materials. This coating not only ensures low frictional resistance but also provides excellent protection against abrasive wear [15, 16]. Among a range of other theoretical interpretations of dry friction processes, also in this case, the previously cited Bowden friction model seems appropriate for a simple and clear description of the phenomena occurring in loaded friction contacts. Therefore, it can be a useful tool in the design and optimization processes for friction pairs in various

Fig. 1.6 Schematic of a design method for reducing dry friction resistance based on Bowden and Tabor theory [4]

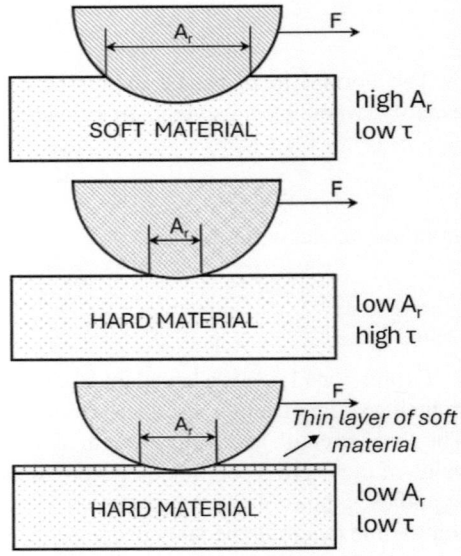

technical and biomedical applications utilizing diamond-like carbon (DLC) coatings. The tribological properties of these coatings, the methods used for their study, and their application possibilities constitute the main focus of this work.

The discussion of the sources of friction, their qualitative and quantitative description, and the possibilities to reduce them, is introduced in the introduction. This is not a coincidence, as it is intended to provide the reader with a simple and understandable idea of the origins of the research on low-friction technical coatings and the ideology behind it. It is acknowledged that this exposition is not exhaustive; the intricacies of contact and friction mechanics are considerably more complex. Readers interested in more comprehensive solutions and explanations of the associations and relationships between contact mechanics and friction are directed to [1, 9].

1.4 Basic Friction and Wear Mechanisms of Coatings

We already know that the primary components of frictional resistance in friction junctions are the adhesive component and the plastic deformation component (plowing). In both isotropic and anisotropic structural compositions at the macroscopic scale, the mechanisms of frictional resistance do not differ from one another. Holmberg et al. proposed another component arising from the resistance resulting from the continuous elastic deformation of one of the surfaces subjected to a normal force and in motion [17]. A schematic illustration of these three components is presented in Fig. 1.7. It should be noted that the proposed fundamental friction mechanisms do not account for surface wear of the interacting friction pair. The wear products

1.4 Basic Friction and Wear Mechanisms of Coatings

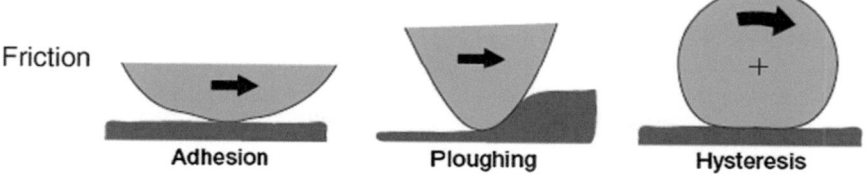

Fig. 1.7 A schematic illustration of friction force components. Reprinted with permission from [17]

involved in the tribological process would undoubtedly make the contact mechanism with their participation significantly more complicated. Nevertheless, the friction force would still arise from the three components presented above.

Wear is the process of material removal from one of the surfaces interacting in the friction junction. Unlike friction, which is instantaneous, wear occurs as a result of the duration during which the interacting surfaces are in contact and move relative to each other. In the context of wear of materials due to friction processes, there are several classifications extensively described in the literature [1–3, 5]. Until recently, wear was classified based on the appearance of the surfaces of the friction pair after the contact process. Examples of such wear mechanisms include various types of scratches, grooves, flaking, pitting, fretting, and scuffing. Most of these pertain to specific applications, such as gears, cam mechanisms, rolling bearing races, or surfaces of sliding bearing bushings. The currently used classification is based on the fundamental mechanisms of material removal and the growing understanding of the basic wear processes. It applies to both coatings and the surface layer itself. A schematic illustration of the wear processes of coatings is presented in Fig. 1.8.

The term cracking generally refers to the destruction of brittle materials. For coatings, it is understood more broadly as a process that begins with the loss of cohesion at the level of the chemical bonds of the material and continues into the phase of propagation and spread of the crack network, inevitably leading to the release of fragments of the coating.

The most commonly used classification includes wear mechanisms for coatings such as: adhesive wear, abrasive wear, fatigue wear, and tribochemical wear [18].

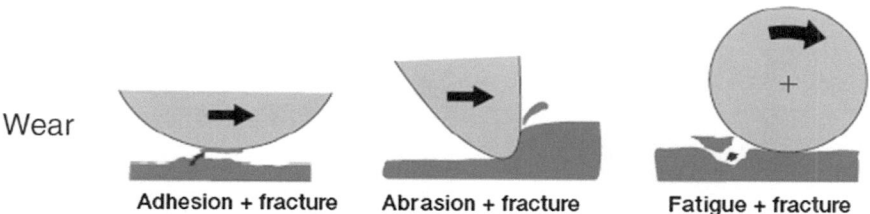

Fig. 1.8 Schematic illustration of coatings wear processes classification. Reprinted with permission from [17]

Adhesive wear—the destruction of the coating surface as a result of the formation and rupture of adhesive bonds, manifested through the creation of micro-junctions and micro-welds between the asperities of the interacting surfaces. The tensile and shear stresses occurring at the junctions, exceeding the material's strength, lead to the formation and propagation of cracks. The spreading crack network results in the detachment of wear products and their transfer to the counterbody.

Abrasive wear—occurs when a harder surface interacts in a friction contact with a softer surface, causing significant deformation of the latter. In the contact zone, shear stresses arise, and locally exceed the critical values. This results in the formation and propagation of a crack network, leading to the detachment of wear products. The wear products can originate from either of the interacting surfaces. When these products move freely in the friction contact, they form what is known as a "third body," which participates in the tribological process acting as an abrasive. In some cases, when these products contain hard phases such as carbides, nitrides, or oxides, they can significantly intensify the abrasive wear process.

As illustrated in Fig. 1.9a, an instance of abrasive wear on a silicon incorporated carbon coating is evident, exhibiting a conspicuous third-body effect. This is characterised by the presence of substantial quantities of wear products accumulated at the periphery of the wear track as well as the couterbody material embedded in the center. Figure 1.9b presents a sever wear of the coating in the central part of the wear track with visible substrate material. In the upper part, there is a visible network of cracks in a partially decohered fragment of the coating. The analysis of the counterbody (Fig. 1.10) confirmed that the primary cause of the accelerated wear of the coating is the abrasive action of the worn ball material. The magnitude of this phenomenon increases in proportion to the duration of the test.

Hard and superhard coating materials are inherently resistant to abrasive wear; however, structures with high hardness tend to exhibit brittle cracking. Such processes can be observed when the mechanical properties of the substrate and the coating are improperly matched. A soft substrate material, failing to provide solid support for the hard and brittle coating, undergoes elastic deformation, leading to intense cracking of the coating and the detachment of wear products. The surface of the

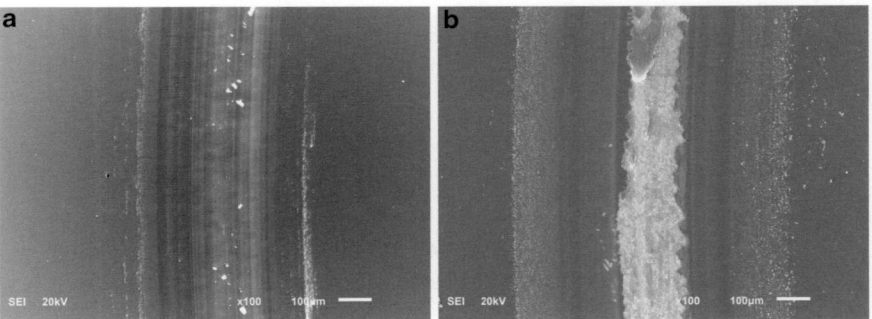

Fig. 1.9 An example of abrasive wear of silicon and oxygen incorporated carbon coating

1.4 Basic Friction and Wear Mechanisms of Coatings

Fig. 1.10 SEM image of 100Cr6 counterbody cooperating with silicon and oxygen incorporated carbon coating

spalling in such cases is quite extensive, and the abrasive wear process can take on a catastrophic nature with significant intensity. An example of this form of wear is shown in Fig. 1.11. The optical microscope image of a crack formed during the adhesion testing of a carbon coating applied to the surface of nitrided steel X53CrMnNi using a diamond indenter revealed a dense network of cracks and areas of complete delamination of coating fragments.

Fatigue wear—is the result of cyclic loading on the surface, under which the material deforms. The generated stresses, primarily of a shear nature, exceed the critical stress values. The cyclic nature of the loads leads to the nucleation and propagation of new cracks, resulting in the spalling of fragments of the coating. In reality, it is commonly believed, the fatigue wear process can be divided into two stages. In the first phase, only surface modification occurs without loss of cohesion and detachment of wear products. As the number of loading cycles increases, the near-surface properties of the coating material gradually change. The second phase begins when the material can no longer sustain the cyclic loads. The result is the nucleation and propagation of cracks, which inevitably lead to material decohesion and catastrophic wear of the coating. This process can sometimes take the form of separating entire fragments of the layer. In the case the damage patches are produced by localised exfoliation of upper coating layers the fatigue wear is considered as cohesive. Debonding along the coating-substrate interface is regarded as an adhesive failure. According to the classical Hertzian theory, maximum contact stresses should be expected at the Bielajew's point located in the area of maximum compressive stresses. However, under tangential loads, typical of sliding contact, this depth decreases, and the point

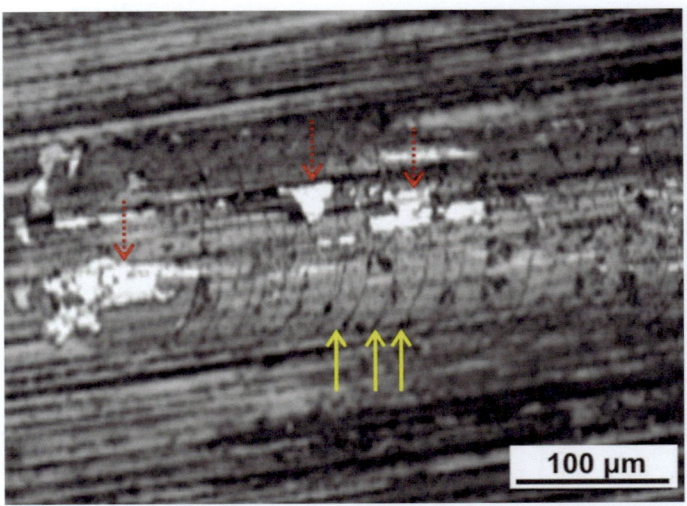

Fig. 1.11 The result of the adhesion test (scratch test) of the carbon coating: the yellow arrows indicate the crack network of the coating formed due to tensile stresses, while the red arrows indicate areas of complete delamination of the coating. Reprinted with permission under Creative Commons BY-NC-ND from [19]

of maximum contact stress shifts closer to the surface. Moreover, the use of hard and wear-resistant coatings in non-conformal applications causes that the dangerous zones are also areas of tensile stress that precede and follow the point of contact (see Fig. 1.12). It is known that these materials are characterized by very high resistance to compression and high sensitivity to the occurrence of tensile stresses. In such a case, the dangerous area is precisely the zone of tensile stresses, which is located at a lower depth in relation to the zone of maximum contact stresses. Taking into account the possibility of structural stress concentration, it can be assumed that there is a high probability of crack nucleation occurring in the tensile stress zone.

In the context of rolling fatigue in coating-substrate systems, the initiation of failure occurs mostly at the interface region. The interfacial failure stress is found to be closely associated with the coating adhesion strength and microstructure, as well as mechanical properties of both the coating and the substrate. In the case of softer substrates and higher loads, the substrate exerts a dual effect on the system. Firstly, it reduces the stress level by means of elastic deformation, and secondly, it increases the stress level in the coating. The result of these effects is the initiation of a crack in the coating or in the interface, with subsequent propagation perpendicular to the surface (see Fig. 1.13a). Increasing the substrate's hardness has been shown to cause cracks, created by the cyclic shear component of the contact stress and propagating in the coating in a parallel direction to the surface (see Fig. 1.13b, c). Mechanisms of crack formation that are perpendicular and parallel to the surface can also occur at the same time. It is characterised by the rapid propagation of initial fine cracks into macroscopic cracks. This expansion is driven by both the extension of existing cracks

1.4 Basic Friction and Wear Mechanisms of Coatings

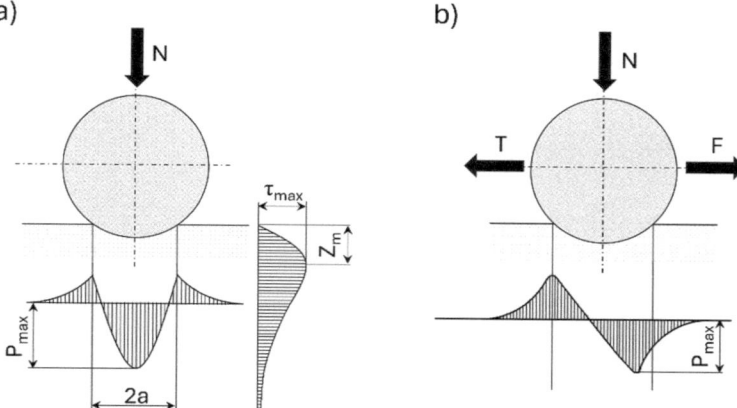

Fig. 1.12 Non-conformal contact characteristics for sphere on flat configuration: **a** under static load and **b** under tangential force; where: a—contact area radius, P_{max}—max contact stress, τ_{max}—max shear stress, Z_m—depth of Bielajew's point

and the coalescence with neighbouring cracks as stress cycling persists. The ultimate outcome of this process is the release of a minute quantity of coating material, which leads to the formation of what is termed "spalling failure" (see Fig. 1.13d). The magnitude and extent of these spots exhibit an increase in proportion to the degree of stress applied [20, 21].

The microstructure of PVD coatings is typically columnar, with the columns becoming coarser as the coating thickness increases. Even in a dense, high-quality PVD TiN coating, the interfaces between columns are likely to be weak links in the coating structure. Consequently, for coatings with a thickness greater than 3 μm, the underlying cause of failure was attributed to defects within the coating microstructure. In the context of rolling contact fatigue testing of PVD coatings, it has been demonstrated that the optimal thickness for TiN and DLC coatings is approximately 0.75 μm [22].

Since the above classification of primary coating wear mechanisms focuses on the method of material removal, ***tribochemical wear*** is not considered an additional mechanism of coating destruction. Chemical processes occurring on the coating surface are important and undoubtedly play a significant role in shaping its functional properties, however, they are not mechanisms that cause material destruction. Due to the reactions taking place on the surface, a surface modification of the material occurs. The nature of these reactions can have both positive and negative impacts on overall functional parameters, such as strength or corrosion resistance. Nonetheless, the fundamental mechanisms of material removal in tribochemical wear processes will follow those listed above.

A separate issue involves coating delamination processes, which are directly related to their mechanical properties (mainly hardness and residual stress), as well as structural factors (thickness and the nature of the interphase with the substrate

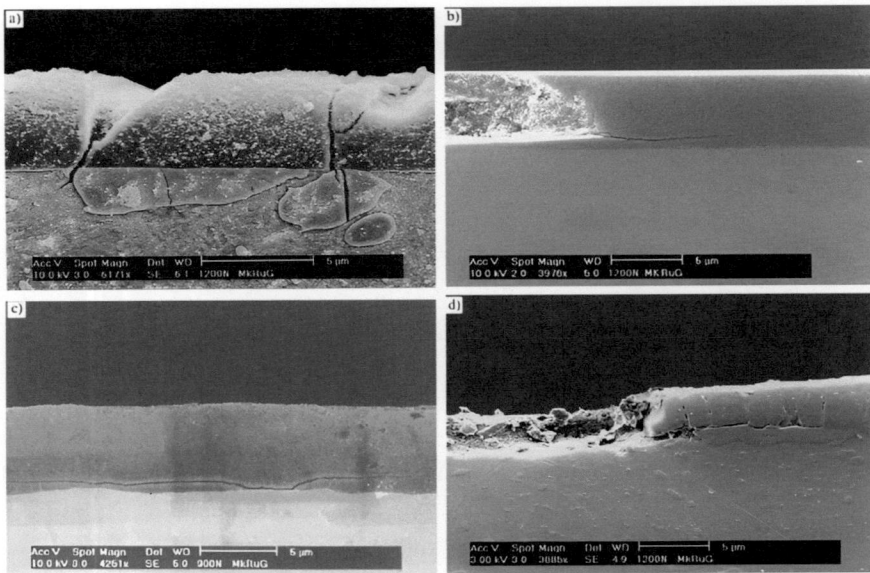

Fig. 1.13 Cross-sectional SEM micrographs showing the fatigue failure modes of TiN coatings deposited by PVD method. Reprinted with permission from [20]

material). The tribological response of the substrate-coating system, resulting from the forces and stresses associated with the characteristics of the friction contact (whether it involves rolling friction, sliding friction, or abrasive action from a rough and hard counterbody), is a combination of the responses from both the coating and the substrate material. If these conditions become sufficiently severe, delamination (meaning the complete separation of the coating from the substrate material), becomes more likely than progressive surface wear. The process typically takes place along the interfacial boundary. Whether delamination occurs at the phase boundary depends on the adhesion forces between the substrate surface and the coating material, load, and other working conditions, as well as the coating thickness, which generally affects residual stress levels. In the case of diffusion bonds, as found in high-temperature chemical vapor deposition (CVD) processes, the issue of coating adhesion and the risk of delamination are virtually nonexistent [3, 23, 24]. For coating systems with a distinct interfacial boundary, measuring adhesion forces is challenging both in terms of implementation and result interpretation, as the outcomes often prove ambiguous. Regions in close proximity to the interfacial boundary may exhibit entirely different property spectra under varying load conditions (tensile or shear). The strength of the bond between the coating and substrate is often described either in terms of normal or transverse destructive forces, i.e., tensile or shear strength (σ_i), or surface energy (G_i), which represents the change in free energy needed to increase the surface area by one unit [10]. Therefore, in most cases

1.4 Basic Friction and Wear Mechanisms of Coatings

of technology used for the synthesis of technical anti-wear coatings, a number of technical procedures are used to improve their adhesion. These include:

- Cleaning or etching the substrate surface prior to deposition, used in galvanic and vacuum technologies
- Ion etching and use of interlayers with high chemical affinity to both components of the coating system, as well as modification of synthesis process parameters to intentionally create a gradient in chemical composition and mechanical properties, used in classical and physically-assisted PVD and CVD processes
- Mechanical surfacre tratments like sand blasting of shot peening, commonly used in thermal spray technologies.

In general terms, coating synthesis technologies, particularly their energetic characteristics, lead to the generation of residual stresses in coating systems, the main sources of which are [18, 25, 26]:

- stresses resulting from the characteristics of the synthesis process (bombardment of the substrate surface and the deposited coating by high-energy particles, chemical reactions and phase transformations)
- thermal stresses arising from the differences in physical properties between the coating and substrate material (thermal expansion coefficient, Young's modulus, Poisson's ratio, thermal conductivity, thickness) as well as the temperature change history during both heating and cooling processes.

Figure 1.14 schematically presents the types of defects found in coatings produced by physical vapor deposition (PVD) methods, whose form and intensity are directly related to the state of residual stresses. Thermal stresses resulting from the different thermal expansion coefficients of the coating and substrate materials can be either compressive or tensile. Tensile stresses may initiate the formation of microcracks through the entire thickness of the coating, which then propagate, creating large cracks that cause the coating to peel off. Compressive stresses promote the propagation of microcracks in the interfacial boundary area, resulting in complete or partial delamination of the coating. Shear stresses occurring between the substrate material and the coating material can also intensify delamination. Coating buckling typically occurs as a result of the synergistic effect of shear and compressive stresses at the interfacial boundary, where the stress value exceeds the critical strength of the coating. In Fig. 1.15 is presented a SEM view of partial, spontaneous delamination of diamond-like carbon coating deposited on steel substrate with use of Ti–C adhesion improving gradient interlayer. Despite the stress releasing effect of the gradient interlayer, a high compressive stress of DLC caused its self-delamination. Moreover, in the front part the whole coating system is delaminated and in the back part delamination occurred at DLC—Ti–C interphase.

Typically, the stress distribution in coatings is not uniform. The presence of stress fields or gradients makes their measurement and analysis even more challenging. However, if we assume that the coating thickness is incomparably smaller than the thickness of the substrate, the stress distribution in the coating is uniform across its entire cross-section and that the stress state in the coating is biaxial, then the strain

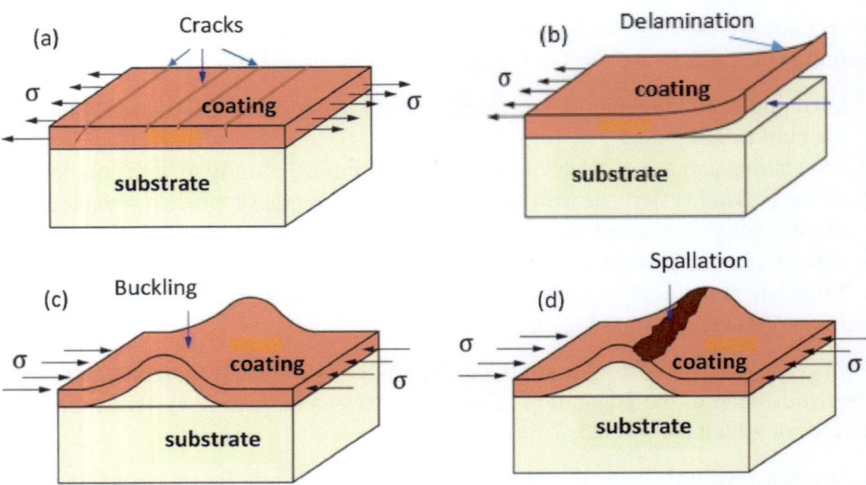

Fig. 1.14 Types of defects found in coatings produced by PVD methods: **a** cracking, **b** delamination, **c** buckling, **d** spallation. Reprinted with permission from [25]

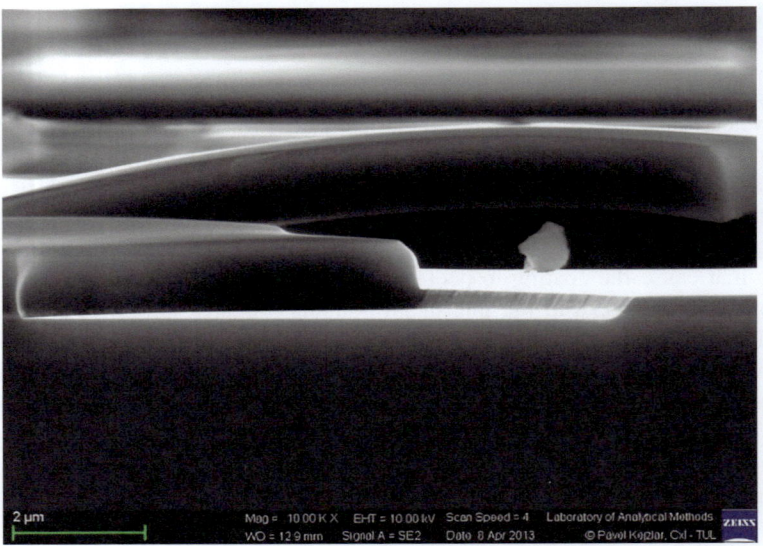

Fig. 1.15 Image of spontaneous partial delamination of DLC coating produced on a steel substrate

energy density Λ_c (the differential of elastic stress and strain) can be determined using the following relation:

$$\Lambda_c = \frac{\sigma_c^2}{E_c}(1 - v_c) \tag{1.19}$$

where σ_c is the value of residual stress, while E_c i v_c are the modulus of elasticity and Poisson's ratio of the coating material, respectively.

The strain energy released per unit area during the delamination of the coating from the substrate is the product of the strain energy density Λ_c and the thickness of the coating. If this product exceeds the interfacial surface energy (G_i), the delamination process will be energetically favored and, once initiated, will continue until complete delamination occurs. If we assume that: (i) the level of residual stress is independent of the coating thickness, (ii) the amount of released strain energy is proportional to the coating thickness, and (iii) the interfacial surface energy is independent of the coating thickness, it can be concluded that thicker coatings will show a greater tendency toward delamination, which is consistent with most literature reports [27–30].

Therefore, in the design of machine components and devices intended for surface modification with anti-wear coatings, application of the criterion of estimating the coating durability based on its wear rate from ongoing classical wear processes, seems overly optimistic. Due to the highly probable risk of unexpected catastrophic failure of the coating due to delamination, the potential of surface engineering technologies to improve the performance of manufactured components is not fully exploited in many industrial sectors.

1.5 Coatings Tribology at Scale

In the processes of modelling, optimisation and prediction of friction and wear in various contact issues, it is particularly important to understand the nature of the basic mechanisms and accompanying phenomena underlying the tribology of friction nodes. Significant contributions to the development of coating-based friction contact design and optimisation issues have been made by the work of Leyland, Matthews and Holmberg [17, 18, 31–34]. In many cases, the mechanisms in question do not occur in pure form; on the contrary, their occurrence is combined and becomes more complex the more complex the contact geometry is. This, in turn, depends on the surface geometrical structure and the release of wear products resulting from the inhomogeneities and time-varying properties of the mating surfaces, as well as the dynamic and kinematic characteristics of the friction node. The basic parameters with a significant impact on friction, wear and varying contact conditions are shown in Fig. 1.16. At first glance, these consist of geometrical, material and energy parameters. However, many of them can also be described at the macro, micro and nano scale level.

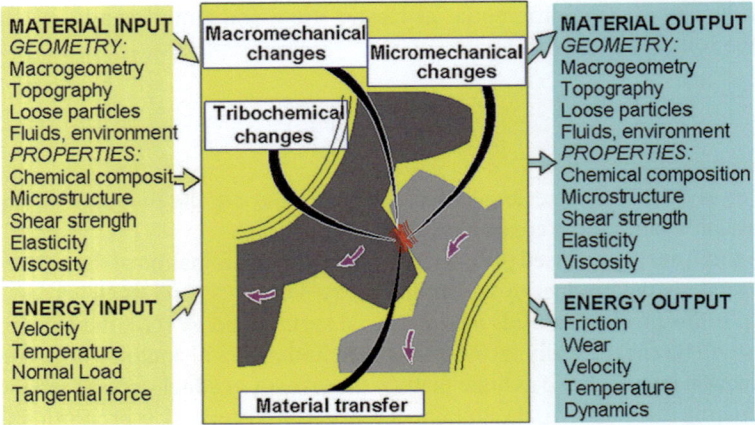

Fig. 1.16 Basic parameters influencing the tribological process. Reprinted with permission from [17]

During the operation of the friction contact, some parameters change:

- surface layers are formed as a result of the reaction of the material surface with its surroundings,
- contacting and plastically deformed surface roughness tips are strengthened due to plastic work
- the adhesion bonds formed are subjected to shearing processes, occurring at some distance from the originally formed contact surface in the softer material outside the zone of plastic deformation and strengthening, and the separated fragments of the softer material are transferred to the counterbody surface
- as a result of insufficient heat dissipation, the temperature in the friction contact may rise locally, causing changes in structure and mechanical properties.

The aforementioned phenomena that can occur in a frictional association mean that, from cycle to cycle, the material and energy characteristics can change significantly and ultimately lead to a new set of parameters affecting wear and friction.

Let us therefore look at tribological phenomena from the point of view of scale. The application of tribological research, conducted at the macro- to nanometric scale using tribometers or microscopes, to the components of machines or devices that are both large and sophisticated, is a crucial and indispensable element in view of the increasing demand for improving the reliability and accessibility of production systems, transport, and public health, both from an industrial and social point of view.

Figure 1.17 shows the different approaches in terms of the scale of identification and understanding of characteristic tribology phenomena over the years.

The increased interest in friction and wear processes and the growing awareness of their importance for the efficiency and reliability of machinery and equipment began during the industrial revolution. Gradually, more and more attention began to be paid

1.5 Coatings Tribology at Scale

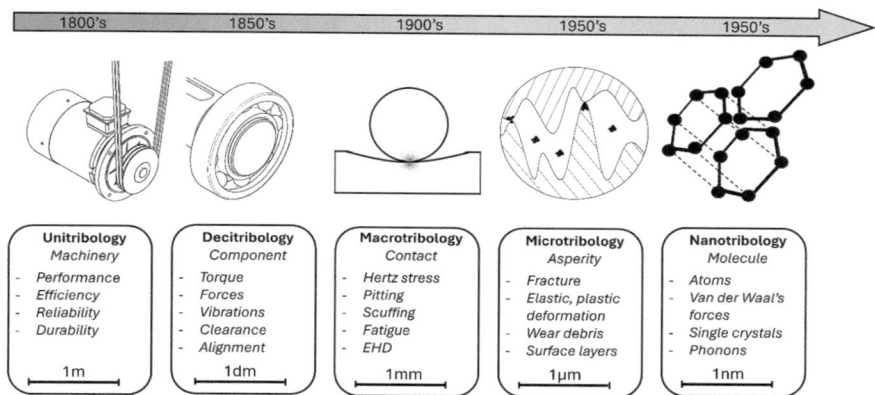

Fig. 1.17 Timeline illustrating the development of tribology in terms of the scale of friction and wear phenomena [33]

to the individual components of friction nodes and their interactions. In the middle of the twentieth century, the scientific discipline we now know as tribology saw a significant development in research, focusing primarily on modelling and studying friction contacts at the level of roughness asperities. Today's tribology is moving towards smaller and smaller dimensions of the phenomena analysed. Undoubtedly, this makes it part of a nano-revolution benefiting greatly from the development of modern research techniques such as atomic force microscopy, nanoindentation or transmission electron microscopy, which enable the analysis of friction and wear phenomena at the molecular and atomic level. This provides extensive opportunities to understand and explain many tribological effects observed at larger scales, while also being a tool of immense importance and cognitive potential. Matthews and Holmberg proposed five scale ranges in tribology [17, 33, 35, 36]:

- nanotribology or molecular tribology covering issues related to interactions between molecules and atoms, such as van der Waals interactions or related phenomena directly determined by the crystallographic structure and binding energy of materials.
- microtribology focusing primarily on the interactions of friction pair elements at the level of roughness asperities. The scope of this scale mainly includes fracture, elastic and plastic deformation processes, surface topography and the formation of surface layers resulting from frictional processes or chemical reactions occurring on the surface.
- macrotribology relating to aspects of stresses occurring at the friction contact interface, the combined load response of the substrate and coating particularly in highly stressed applications such as gears or bearing components. These stresses are responsible for the wear mechanisms observed in friction contacts, such as galling, grooving, pitting or peeling of the coating due to low adhesion to the substrate surface.

- decytribology, which deals with the definition and measurement of typical operating parameters of friction nodes resulting from the interaction of their components, such as friction torque, forces occurring at the contact, vibrations (oscillations), fits or alignment.
- unitribology encompassing all tribological factors that affect the overall performance and efficiency of machinery and equipment, not excluding their statistical reliability.

Figure 1.18 shows typical contact conditions occurring at the macroscale for the case of a high hardness sphere interacting with a surface modified with a thin anti-wear coating. Even though the number of parameters that can influence the characteristics of the tibological process is large, there is still a real possibility to control it. Their proper identification and an understanding of the relationships between them is a key element in the discussion of the correct prediction and controllability of friction and wear processes. In this case, the parameters of particular importance are the hardness of the substrate and the coating, as well as their relationship to each other (hard substrate and soft coating or soft substrate and hard coating), the thickness of the coating, the surface roughness and the fragments of removed material actively involved in the tribological process.

Each of the parameters listed is equally important individually but also critical in terms of their correlation with each other. The picture becomes much more complicated when we look at it from a scale perspective. Namely, the occurrence of many phenomena accompanying friction processes goes well beyond the macro or micro scale adopted in this case. It can often be seen that mechanisms on a nano-scale

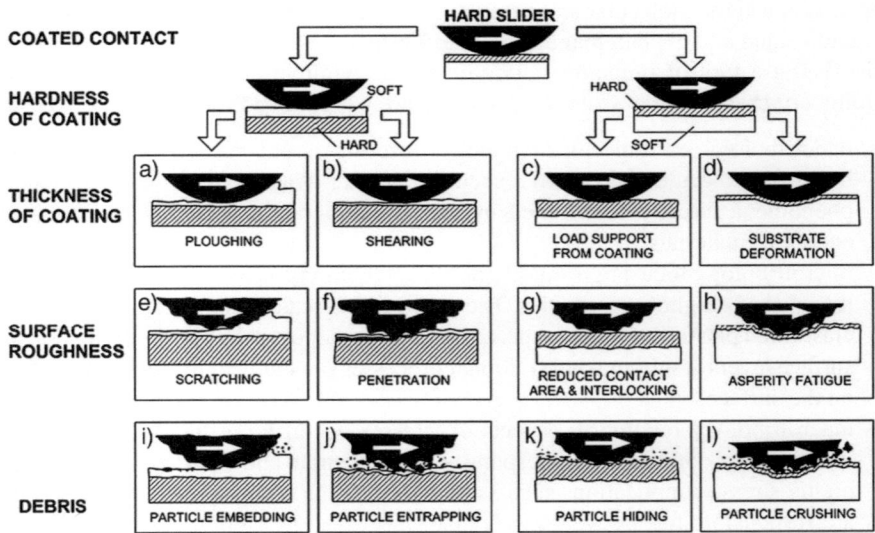

Fig. 1.18 Main parameters affecting friction on a macro scale [17]

influence phenomena on a micro-scale and, conversely, the causes of friction and wear on a macro-scale can be traced back to mechanisms on a micro-scale. Thus, the full picture of friction and wear phenomena occurring in friction nodes of coating systems is a synergic effect of the occurrence of a number of tribological mechanisms observed mainly in the area of the first three scale levels, i.e. from nano to macro. Since within this scale range the coating selection, its performance and influence of the overall tribological characteristics of the friction contact appears to be the most actively investigated, we shall discuss some aspects related to mechanisms at each of these scales.

1.5.1 Nanoscale Mechanisms

The advent of new technologies, coupled with the development of innovative research methods and techniques, has led to a deeper understanding of the underlying mechanisms of friction at the nanoscale. This has been made possible by the collaboration between tribologists and surface engineers. The rapid advancement of technology has facilitated the emergence of novel research techniques, allowing for the analysis of friction and wear phenomena at the molecular level and the investigation of friction forces between contacting molecules at the nano-Newton scale. Once more, in accordance with the established theory of friction proposed by Bowden and Tabor [4] the friction force was predominantly ascribed to the minute irregularities of the contacting surfaces. The surface asperities make contact and exert a force of pressure on one another, resulting in the formation of regions with an area that is directly proportional to the magnitude of the friction force. This, in turn, may suggest that the phenomenon of friction arises from the formation of sufficiently strong bonds at the points of true contact, resulting in the tearing away of small fragments of the material. Tabor and his student, by developing an apparatus for measuring atomic-scale friction, demonstrated that this hypothesis was erroneous. They presented an example of friction occurring in the absence of wear for the first time [37]. This, in turn, revealed that an alternative mechanism was responsible for a significant proportion of the energy dissipation observed at the macroscopic scale. It is already established that the atoms which comprise a crystalline network are not rigidly bonded and may oscillate around their equilibrium positions. If two crystalline surfaces slide over one another, atoms in close proximity to the interface may be set into motion by the sliding action of atoms on the opposing surface. This results in the propagation of elastic waves outward, carrying away energy from the contact area. Such lattice disturbances are referred to as phonons. The quartz crystal microbalance (QCM), initially employed for microweighing and time standard purposes, is an instrument operating at a frequency sufficient to detect phonons with a lifetime of no more than a few tens of nanoseconds. The atomic vibrations generated in friction contact, initially measured by QCM, were subsequently attributed to a phononic energy dissipation mechanism and identified as an additional source of friction [38].

Further advances in nanoscale mechanisms analysis were made possible by techniques employed by surface engineers for the development and structural characterisation of nanometre-scale interfaces. Ultrahigh vacuum technology enabled the synthesis of well-defined crystalline surfaces, designed for friction investigations, which usually involved contact at one instead of multiple asperities.

The development of research methods and techniques has enabled significant progress to be made in understanding and analysing the complex nanophysical phenomena that occur in the contact conditions of co-operating surfaces. This has opened up a wide range of research opportunities, both in terms of the friction and wear mechanisms themselves, and in the study of the surface topography, chemical structure or chemical and phase composition of the mating surfaces. The results of these studies have provided new insights into the selection and optimisation of modified surfaces with respect to their intended use or operating conditions and environment. The tool that has revolutionised this research has undoubtedly been atomic force microscopy (AFM). This technique is one of the most important tools for imaging, measuring and manipulating matter at the nanoscale. At its core, it is commonly used to analyse surface topography with a resolution of fractions of a nanometre. Precision piezoelectric elements are responsible for accurately scanning the surface of the material under investigation, and the displacement of the mechanical scanning probe is measured using a laser and photodiode array. However, thanks to the possibility of measuring the lateral force exerted on the scanning probe in contact with the analysed surface, it is also possible to determine its tribological properties for loads ranging from one to several tens of nN. The frictional force between the scanning probe and the analysed surface causes a lateral torsion of the probe which is recorded by the optical system and which, in relation to the load and the mechanical properties of the probe, allows the determination of the coefficient of friction or to obtain an image of the so-called friction contrast for nanocomposite materials [38–41]. The adaptation of the atomic force microscope to the study of friction in ultra-high vacuum has opened a new door to gain a better insight into very important and fundamental questions about friction that have been raised in recent years, mainly from the point of view of the energy dissipation mechanisms of friction processes, as well as a better understanding of the influence of adsorbed chemical compounds on the frictional properties of surfaces [42].

However, even with careful calibration, AFM experiments do not provide a complete picture of the individual forces between the scanning probe and the surface atom. Computational techniques, including Molecular Dynamics (MD), provide information about each atom present at the interface between the interacting phases, and are therefore complementary to and uniquely suited for use with experimental techniques. The dynamic development of numerical methods, together with the increasing computing power of supercomputers, provides extensive opportunities for theoretical studies of friction between solids. Molecular dynamics simulation methods are based on modelling approximations to interatomic forces, the analysis of which allows contact mechanisms to be studied at the atomic scale [43]. Molecular dynamics simulations were used to investigate the mechanical and tribological properties of amorphous carbon thin films with and without surface hydrogen. The

1.5 Coatings Tribology at Scale

simulations showed that the three-dimensional structure, and not just the sp^2/sp^3 hybridised carbon ratio, is crucial in determining the mechanical properties of the films [44].

Undoubtedly, an important element in research of nanometer scale friction and wear mechanisms are methods enabling probing of external vibrational modes. Namely, Raman Spectroscopy [15, 45, 46], Energy Electron Loss Spectroscopy (EELS) [47] and Fourier Transform Infrared Spectroscopy [48, 49]. However, these techniques will be discussed in more detail in later chapters of this book with reference to specific cases of friction and wear analysis of carbon based coatings.

1.5.2 Microscale Mechanisms

The basic tribological mechanisms observed at the micro level include stress and strain problems at the level of roughness peaks, crack propagation or the release of wear products. The main mechanisms identified for crack nucleation and propagation leading to progressive wear of the coating are shear and fracture processes, but despite many hypotheses, these fundamental phenomena are still not fully understood and remain in the realm of research [50–52]. A more unified approach to friction and wear at the microscale appears to be the third body mechanism, the process of which is shown in Fig. 1.19. It involves the gradual removal of coating material, which is then transferred to the counterpart surface. During this time, the coating surface and the resulting transition layer react chemically with the surrounding atmosphere to form new compounds, which are identified from a phase composition point of view. As the thickness of the transition layer increases, fragments of it are removed from the surface by shear forces and then enter or are removed from the contact zone.

The resulting transition layer forms in the near-surface contact areas of the surface roughness peaks as a result of chemical reactions or material phase transformations induced by high contact stresses, local temperature gradient fields or so-called flash temperatures at the points of thermal pulse generation. This layer usually has the character of micro-areas with low shear strength. It is therefore important that they are present in sufficient quantity to ensure a low coefficient of friction, while significantly reducing the risk of rapid surface wear. In some cases, wear products trapped in the contact zone can produce positive results. By absorbing the speed differences of the

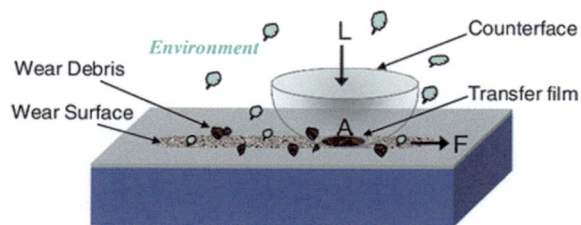

Fig. 1.19 Schematic of detachment and transfer process that produce third bodies during sliding in air. Reprinted with permission from [53]

mating surfaces and the energy of deformation, they reduce the wear of the friction node components [54]. On the other hand, the transition layer formed on the mating surface can be mechanically or chemically enriched during the tribochemical process by wear products of the coating or the counterbody material agglomerating on its surface. This can lead to deterioration of the mating surfaces or progressive spalling of fragments of the transition layer, resulting in an increase in the coefficient of friction and the rate of wear of the friction contact elements [54, 55]. Figures 1.20 and 1.21 show a graph illustrating the abrupt changes in the coefficient of friction due to the accumulation and detachment of fragments of the transition layer from the counterbody surface, and a SEM image of the counterbody with a transition layer containing numerous cracks and areas of spalling.

These processes are underpinned by a range of mechanisms, including elastic deformation, fracture, shear and rolling. Therefore, from the point of view of analysing friction and wear processes of coating materials at the microscale, the important parameters characterising the tribological properties of a friction node appear to be the elastic modulus, shear strength and fracture toughness of both the coating material and the substrate [34].

In this respect, recent works on the synthesis and properties of protective technical coatings clearly demonstrate the growing importance of nanocomposite materials. In particular, the small grain sizes of nanocomposite coatings in the sub-10 nm range make them a new generation of materials with unique performance characteristics, often completely new compared to conventional coating materials with grain sizes above 100 nm [56]. Figure 1.22 shows the 4 main types of nanostructures used in today's hard and wear resistant coating applications. These include multilayer nanocomposite coatings (bilayers with a recurrence period of the order of nm), columnar nanostructures, nanograins surrounded by a very thin coating of amorphous matrix, and nanostructured coatings in the form of a mixture of nanograins

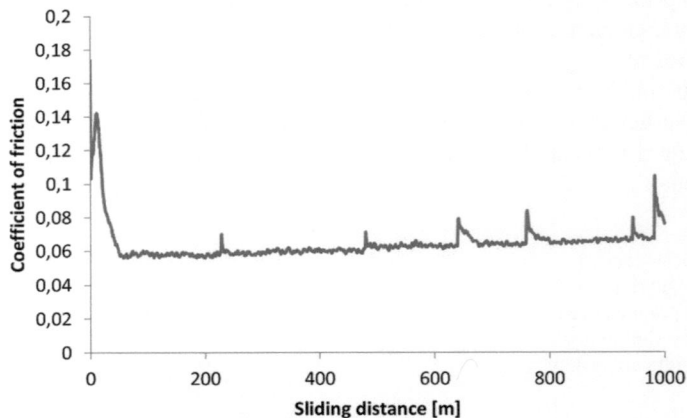

Fig. 1.20 Evolution of coefficient of friction of a-C:H:SiO$_x$ layer with visible abrupt changes caused by third body removal effect

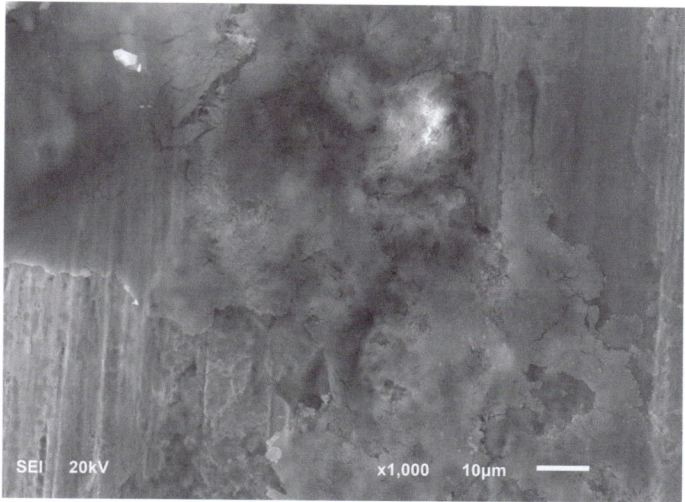

Fig. 1.21 Surface of the AISI 316L counterbody covered with the transfer layer with visible cracks and places of delamination

with different crystallographic orientations. Individual nanostructures are produced with different process parameters, either by synthesising successive interlayers or by modifying the nanostructure in so-called transition regions, where it changes from crystalline to nanocrystalline, and finally to amorphous [57].

The significant increase in hardness of the coatings (up to twice the hardness of the harder component of the nanocomposite) is caused by the interactions of the matrix boundary regions with the individual grains surrounding them, as well as interactions between the individual grains. Furthermore, the direct interactions between the individual grains (with different crystallographic orientations) are stronger the

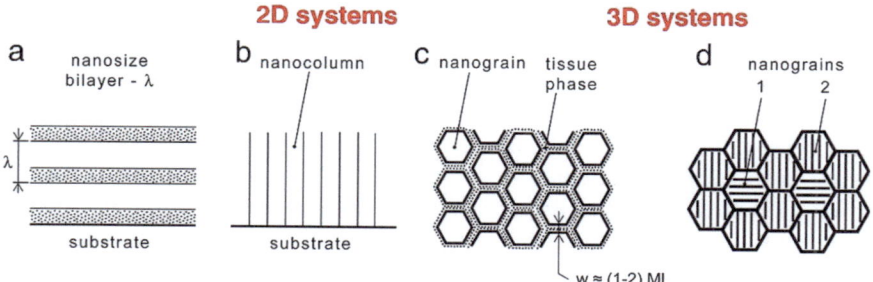

Fig. 1.22 Schematic illustration of four types of nanostructures of nanocomposite coatings with improved hardness and abrasion resistance **a** bilayer with a repeatability period of nm; **b** columnar nanostructure; **c** nanorods surrounded by a matrix phase; **d** mixture of nanograins with different crystallographic orientation [57]

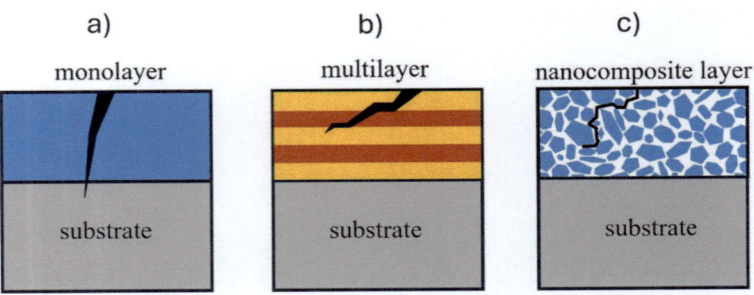

Fig. 1.23 Fracture resistance. Schematic of crack propagation in: **a** a monolithic coating; **b** a multilayer coating with a predetermined recurrence period; and **c** a nanomoposite coating in the form of nanocrystalline grains in an amorphous matrix

smaller the size of the 'nanoparticles' and are most pronounced for grain sizes below 10 nm. In such an arrangement, the number of atoms forming the individual grains is comparable or even smaller than in the surrounding matrix boundary regions. Such a structure primarily inhibits the formation and movement of dislocations. Figure 1.23 shows schematically the crack propagation path for the three types of coatings.

In a single coating, the crack propagation path is almost seamless, perpendicular to the surface. The successive cracks that occur, form a network, resulting in rapid release of coating fragments and thus progressive wear or delamination. In the case of a multilayer coating, the crack propagation process can be slowed down or sometimes stopped at the interface between the individual layers. In the case of a nanocomposite coating, the crack can propagate only along the soft matrix. In both cases, the strengthening of the grain boundaries and the blocking of their mutual displacement prevents crack propagation and also significantly reduces the residual stress value of nanocomposite coatings [58, 59].

1.5.3 Macroscale Mechanisms

At the macro level, the mechanical response of the system in terms of stress distribution and elastic and plastic deformation across the contact zone is important from the point of view of tribological mechanisms. With these considerations of the tribological mechanisms observed at the nano- and micro-scales in mind, it is worth returning to Fig. 1.19 and re-examining the main factors influencing friction at the macroscale. The parameters that are of primary significance in this context are as follows: (i) the hardness and thickness of the coating, (ii) the parameters of its surface geometrical structure, and (iii) the wear products generated by friction. It is evident that the thickness of the coatings, the parameters of their surface geometrical structure, and the size of the material fragments released are of the same order of magnitude, varying from hundredths of a part to a few microns. However, it is these factors, or more precisely their relationship to each other, that will exert a decisive influence on the friction and wear processes for each individual case considered on a macro scale. Hardness, however, should not be overlooked as a fundamental parameter in the characterisation of technical coatings that perform anti-friction and wear protection functions. As mentioned above, the interrelationship between the hardness of the substrate and the coating is an important factor affecting the load carrying capacity of the tribological system and also determines the likelihood of cracking or delamination of the coating due to elastic or plastic deformation of the substrate. Contact conditions in terms of tangential and normal loads are critical in shaping the tribological characteristics of a friction node from the point of view of stress and strain as well as friction and wear. Therefore, in addition to hardness as a parameter characterising both the substrate and the coating, their elasticity is equally important. There is no simple relationship between the mechanical response of a coating and the hardness or modulus values themselves, but these properties are strongly dependent on the ratio of hardness to Young's modulus (H/E). The importance of the H/E ratio in determining the strength of a coating, in particular its ability to accommodate deformation of the substrate under applied loads and thus its resistance to mechanical wear and cracking, was demonstrated by Leyland and Matthews [60]. Musil showed that nanocomposite coatings of similar hardness but different chemical composition have different values of effective Young's modulus and thus different elastic recovery (We). This parameter is proportional to the H^3/E^2 ratio and characterises the plastic deformation resistance of the coating [61]. The combination of elevated hardness and reduced modulus of elasticity has been demonstrated to yield enhanced wear resistance and prolonged coating durability. Consequently, this enables the mechanical properties of the coating material to be aligned with the requirements of the specific application [62, 63]. Monolithic hard coatings of considerable thickness are known for their high load carrying capacity. However, under the assumption of elastic or plastic deformation of the substrate, there is a susceptibility to bending stresses, which can result in the initiation and propagation of cracks. Flexible nanocomposite coatings, which are produced by an alternative synthesis of interlayers that exhibit distinct differences in hardness, appear to be a significantly more effective solution. Such coating systems represent a new class of hard and flexible materials with enhanced fracture toughness, characterised by (i) a low effective Young's modulus resulting in an H/E ratio ≥ 0.1 and $W_e \geq 60\%$; (ii) a dense and void-free microstructure and (iii) a compressive stress state [64]. The key to the design and manufacture of such systems lies in the choice of technology and the optimisation of synthesis parameters to ensure the production of coatings that meet all the above criteria. The simultaneous effect of these parameters on a significant improvement in fracture toughness has been demonstrated, for example, for oxide/oxide coating systems (Zr–Al–O) [65], oxide/nitride (Al–O–N) [66] and nitride/nitride (Ti–Ni–N) or Al–Cu–N [67, 68]. Figure 1.24 shows SEM images of the surface morphology of Cr–Cu–O and Al–Cu–N coatings subjected to bending on a cylinder surface with radius r = 10 mm. The Cr–Cu–O coating was characterised by low values of H/E ratio and elastic recovery, while for the Al–Cu–N coating the values of these parameters met the above criteria.

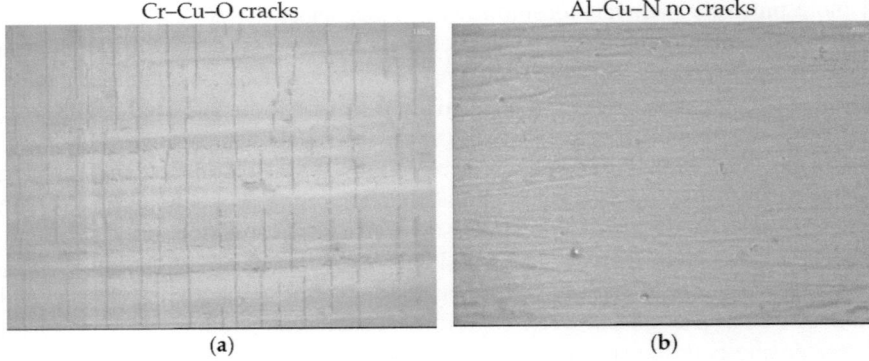

Fig. 1.24 SEM image of the surface morphology of **a** Cr–Cu–O coating **b** Al–Cu–N coating. Reprinted with permission under Creative Commons Attribution CC BY from [68]

It is important to note that this group of nanocomposite coatings, which combine the advantages of both hard and soft phases, can also exhibit additional unique physical and functional properties. A prime example of this is Al–Cu–N hard flexible coatings with antibacterial properties [69].

References

1. Ian, H., Philip, S.: Tribology, Friction and Wear of Engineering Materials (2017)
2. Szczerek, M., Wiśniewski, M.: Tribologia i tribotechnika. SIMP-PTT-ITeE, Radom, Radom (2000)
3. Kula, P.: Inżynieria warstwy wierzchniej. Wydawnictwo Politechniki Łódzkiej (2000)
4. Bowden, F.P., Tabor, D.: Friction and Lubrication of Solids. The Clarendon Press, Oxford (1964)
5. Williams, J.: Engineering Tribology. Cambridge University Press (2005)
6. Johnson, K.L.: Contact Mechanics. Cambridge University Press (1985)
7. Rakowski, W., Rusinek, M.: Naprężenia stykowe przy współpracy kół zębatych pokrytych powłoką. Tribologia. **6**, 173–182 (2011)
8. Yang, W., Bai, P., Fang, J., Qi, K., Zhou, Q.: Impact contact behaviors of elastic coated medium with imperfect interfaces. Int. J. Mech. Sci. **236**, 107743 (2022). https://doi.org/10.1016/j.ijmecsci.2022.107743
9. Popov, V.L.: Contact Mechanics and Friction: Physical Principles and Applications. Springer, Berlin Heidelberg (2010)
10. Płaza, S., Margielewski, L., Celichowski, G.: Wstęp do tribologii i tribochemii. Wydawnictwo Uniwersytetu Łódzkiego (2005)
11. Szczerek, M., Tuszyński, W.: Badania Tribologiczne Zacieranie. Instytut Technologii Eksploatacji, Radom (2000)
12. Mann, A.B.: Nanotribology and Nanomechanics. Springer International Publishing, Cham (2017)
13. Zhang, P., Li, S.X., Zhang, Z.F.: General relationship between strength and hardness. Mater. Sci. Eng. A **529**, 62–73 (2011). https://doi.org/10.1016/J.MSEA.2011.08.061
14. Davis, R.J., (ed.).: Materials for sliding bearings. In: Metals Handbook Desk Edition, pp. 695–701. ASM International (1998)

References

15. Robertson, J.: Diamond-like amorphous carbon. Mater. Sci. Eng. R. Rep. **37**, 129–281 (2002). https://doi.org/10.1016/S0927-796X(02)00005-0
16. Erdemir, A., Donnet, C.: Tribology of diamond-like carbon films: Recent progress and future prospects. J. Phys. D. Appl. Phys. **39** (2006). https://doi.org/10.1088/0022-3727/39/18/R01
17. Holmberg, K., Ronkainen, H., Laukkanen, A., Wallin, K.: Friction and wear of coated surfaces—scales, modelling and simulation of tribomechanisms. Surf. Coat. Technol. **202**, 1034–1049 (2007). https://doi.org/10.1016/j.surfcoat.2007.07.105
18. Holmberg, K., Matthews, A.: Coatings Tribology—Properties, Mechanisms, Techniques and Applications in Surface Engineering. Elsevier (2009)
19. Batory, D., Szymanski, W., Clapa, M.: Mechanical and tribological properties of gradient a-C:H/Ti coatings. Mater. Sci. Poland **31**, 415–423 (2013). https://doi.org/10.2478/s13536-013-0121-9
20. Carvalho, N.J.M., Huis In 't Veld, A.J., De Hosson, J.T.: Interfacial fatigue stress in PVD TiN coated tool steels under rolling contact fatigue conditions. Surf. Coat Technol. **105**, 109–116 (1998)
21. Chang, T.-S.P., Cheng, H.S., Sproul, W.D.: The influence of coating thickness on lubricated rolling contact fatigue life. In: Metallurgical Coatings and Thin Films 1990, pp. 699–708. Elsevier (1990)
22. Stewart, S., Ahmed, R.: Rolling contact fatigue of surface coatings-a review (2002)
23. Choy, K.: Chemical vapour deposition of coatings. Prog. Mater. Sci. **48**, 57–170 (2003)
24. Burakowski, T., Wierzchoń, T.: Inżynieria powierzchni metali. Wydawnictwa Naukowo Techniczne, Warszawa (1995)
25. Riyadi, T.W.B., Setiadhi, D., Anggono, A.D., Siswanto, W.A., Al-Kayiem, H.H.: Analysis of mechanical and thermal stresses due to TiN coating of Fe substrate by physical vapor deposition. Forces Mech. **4**, 100042 (2021). https://doi.org/10.1016/j.finmec.2021.100042
26. Uhlmann, E., Klein, K.: Stress design in hard coatings. Surf. Coat. Technol. **131**, 448–451 (2000). https://doi.org/10.1016/S0257-8972(00)00837-9
27. Heidarinejad, A., Ashrafizadeh, F.: Influence of surface texture and coating thickness on adhesion of nickel plated coatings to aluminium substrate. J. Manuf. Process. **120**, 435–448 (2024). https://doi.org/10.1016/j.jmapro.2024.04.040
28. Zhou, T., Chen, H., Yue, Y-nan, Y., Fang, X.Y., Zhang, R.Q., Gao, X., Cai, Z.B.: Effect of coating thickness on interfacial adhesion and mechanical properties of Cr-coated zircaloy. Trans. Nonferrous Metals Soc. China (English Edition) **33**, 2672–2686 (2023). https://doi.org/10.1016/S1003-6326(23)66289-2
29. Chen, Z., Zhou, K., Lu, X., Lam, Y.C.: A review on the mechanical methods for evaluating coating adhesion (2014)
30. Sheeja, D., Tay, B.K., Leong, K.W., Lee, C.H.: Effect of film thickness on the stress and adhesion of diamond-like carbon coatings (2002)
31. Holmberg, K., Mathews, A.: Coatings tribology: a concept, critical aspects and future directions. Thin Solid Films **253**, 173–178 (1994). https://doi.org/10.1016/0040-6090(94)90315-8
32. Holmberg, K.: Reliability Aspects of Tribology (2001)
33. Matthews, A., Franklin, S., Holmberg, K.: Tribological coatings: contact mechanisms and selection. J. Phys. D Appl. Phys. **40**, 5463–5475 (2007). https://doi.org/10.1088/0022-3727/40/18/S07
34. Holmberg, K., Matthews, A.: Tribological properties of metallic and ceramic coatings. In: Bhushan, B. (ed.) Modern Tribology Handbook. CRC Press (2000)
35. Holmberg, K.: Reliability aspects of tribology. Tribol. Int. **34**, 801–808 (2001). https://doi.org/10.1016/S0301-679X(01)00078-0
36. Holmberg, K., Matthews, A.: Coatings Tribology: Properties Techniques and Applications in Surface Engineering. Elsevier, Amsterdam (1994)
37. Bhushan, B., Israelachvili, J.N., Landman, U.: Nanotribology: friction, wear and lubrication at the atomic scale. Nature **374**, 607–616 (1995). https://doi.org/10.1038/374607a0
38. Krim, J.: Surface science and the atomic-scale origins of friction: what once was old is new again

39. Martínez-Martínez, D., Kolodziejczyk, L., Sánchez-López, J.C., Fernández, A.: Tribological carbon-based coatings: an AFM and LFM study. Surf. Sci. **603**, 973–979 (2009). https://doi.org/10.1016/j.susc.2009.01.043
40. Martínez-Martínez, D., Sánchez-López, J.C., Rojas, T.C., Fernández, A., Eaton, P., Belin, M.: Structural and microtribological studies of Ti–C–N based nanocomposite coatings prepared by reactive sputtering. Thin Solid Films **472**, 64–70 (2005). https://doi.org/10.1016/j.tsf.2004.06.140
41. Koinkar, V.N., Bhushan, B.: Microtribological studies of A1203, AI203-TiC, polycrystalline and single-crystal Mn–Zn ferrite, and SiC head slider materials (1996)
42. Van Den Oetelaar, R.J.A., Flipse, C.F.J.: Atomic-scale friction on diamond (111) studied by ultra-high vacuum atomic force microscopy
43. Harrison, J.A., Gao, G., Schall, J.D., Knippenberg, M.T., Mikulski, P.T.: Friction between solids. Philos. Trans. R. Soc. A Math. Phys. Eng. Sci. **366**, 1469–1495 (2008). https://doi.org/10.1098/rsta.2007.2169
44. Gao, G.T., Mikulski, P.T., Chateauneuf, G.M., Harrison, J.A.: The Effects of Film Structure and Surface Hydrogen on the Properties of Amorphous Carbon Films. https://doi.org/10.1021/jp034544
45. Ferrari, A.C., Robertson, J.: Interpretation of Raman spectra of disordered and amorphous carbon. Phys. Rev. B **61**, 14095–14107 (2000). https://doi.org/10.1103/PhysRevB.61.14095
46. Casiraghi, C., Piazza, F., Ferrari, A.C., Grambole, D., Robertson, J.: Bonding in hydrogenated diamond-like carbon by Raman spectroscopy. In: Diamond and Related Materials, pp. 1098–1102 (2005)
47. Rico, V.J., Yubero, F., Espinós, J.P., Cotrino, J., González-Elipe, A.R., Garg, D., Henry, S.: Determination of the hydrogen content in diamond-like carbon and polymeric thin films by reflection electron energy loss spectroscopy. Diam. Relat. Mater. **16**, 107–111 (2007). https://doi.org/10.1016/j.diamond.2006.04.002
48. Kumar, N., Barve, S.A., Chopade, S.S., Kar, R., Chand, N., Dash, S., Tyagi, A.K., Patil, D.S.: Scratch resistance and tribological properties of SiOx incorporated diamond-like carbon films deposited by r.f. plasma assisted chemical vapor deposition. Tribol. Int. **84**, 124–131 (2015). https://doi.org/10.1016/j.triboint.2014.12.001
49. Jedrzejczak, A., Szymanski, W., Kolodziejczyk, L., Sobczyk-Guzenda, A., Kaczorowski, W., Grabarczyk, J., Niedzielski, P., Kolodziejczyk, A., Batory, D.: Tribological Characteristics of a-C:H:Si and a-C:H:SiOx coatings tested in simulated body fluid and protein environment. Materials **15** (2022). https://doi.org/10.3390/ma15062082
50. Kato, K., Adachi, K.: Wear Mechanisms. Presented at the December 28 (2000)
51. Gerde, E., Marder, M.: Friction and fracture. Nature **413**, 285–288 (2001). https://doi.org/10.1038/35095018
52. Godet, M.: Third-bodies in tribology. Wear **136**, 29–45 (1990). https://doi.org/10.1016/0043-1648(90)90070-Q
53. Scharf, T.W., Prasad, S.V.: Solid lubricants: a review. J. Mater. Sci. **48**, 511–531 (2013). https://doi.org/10.1007/s10853-012-7038-2
54. Singer, I.L.: How third-body processes affect friction and wear. MRS Bull. **23**, 37–40 (1998). https://doi.org/10.1557/S088376940003061X
55. Jedrzejczak, A., Kolodziejczyk, L., Szymanski, W., Piwonski, I., Cichomski, M., Kisielewska, A., Dudek, M., Batory, D.: Friction and wear of a-C:H:SiOx coatings in combination with AISI 316L and ZrO_2 counterbodies. Tribol. Int. **112**, 155–162 (2017). https://doi.org/10.1016/j.triboint.2017.03.026
56. Wendler, B.: Functional coatings by PVD or CVD methods. ITeE-PIB, Lodz (2011)
57. Musil, J.: Hard nanocomposite coatings: thermal stability, oxidation resistance and toughness. Surf. Coat. Technol. **207**, 50–65 (2012). https://doi.org/10.1016/j.surfcoat.2012.05.073
58. Musil, J., Vlček, J.: Magnetron sputtering of hard nanocomposite coatings and their properties. Surf. Coat. Technol. **142–144**, 557–566 (2001). https://doi.org/10.1016/S0257-8972(01)01139-2

References

59. Veprek, S., Reiprich, S.: A concept for the design of novel superhard coatings. Thin Solid Films **268**, 64–71 (1995)
60. Leyland, A., Matthews, A.: On the significance of the H/E ratio in wear control: a nanocomposite coating approach to optimised tribological behaviour (2000)
61. Musil, J., Kunc, F., Zeman, H., Polaková Poláková´, H.: Relationships between hardness, Young's modulus and elastic recovery in hard nanocomposite coatings (2002)
62. Fildes, J.M., Meyers, S.J., Mulligan, C.P., Kilaparti, R.: Evaluation of the wear and abrasion resistance of hard coatings by ball-on-three-disk test methods-a case study. Wear **302**, 1040–1049 (2013). https://doi.org/10.1016/j.wear.2012.11.018
63. Miletić, A., Čekada, M., Kovačević, L., Škorić, B.: Elastic-plastic behavior of hard ceramic coatings. J. Technol. Plast. **42** (2017). https://doi.org/10.24867/jtp.2017.42-2.11-22
64. Musil, J.: Flexible Hard Nanocomposite Coatings (2015)
65. Musil, J., Sklenka, J., Cerstvy, R.: Transparent Zr–Al–O oxide coatings with enhanced resistance to cracking. Surf. Coat. Technol. **206**, 2105–2109 (2012). https://doi.org/10.1016/j.surfcoat.2011.09.035
66. Musil, J., Jílek, R., Meissner, M., Tölg, T., Čerstvý, R.: Two-phase single layer Al–O–N nanocomposite films with enhanced resistance to cracking. Surf. Coat. Technol. **206**, 4230–4234 (2012). https://doi.org/10.1016/j.surfcoat.2012.04.028
67. Musil, J., Jílek, R., Čerstvý, R.: Flexible Ti–Ni–N Thin Films Prepared by Magnetron Sputtering (2014)
68. Musil, J.: Flexible Antibacterial Coatings (2017)
69. Musil, J., Blažek, J., Fajfrlík, K., Čerstvý, R.: Flexible antibacterial Al–Cu–N films. Surf. Coat. Technol. **264**, 114–120 (2015). https://doi.org/10.1016/j.surfcoat.2015.01.006

Chapter 2
Solid Lubrication

Abstract The chapter commences with the establishment of a theoretical foundation for the mitigation of sliding friction resistance. Subsequently, a categorisation of coating materials employed as solid lubricants is presented. The chapter then proceeds to address alternative methods of surface modification that have been demonstrated to result in friction and wear reduction. Ultimately, the chapter culminates in a deliberation on the synergistic effect of these two approaches on the comprehensive enhancement of the tribological properties of friction nodes.

As far back as ancient Egypt, lubricants in the form of oil or grease were introduced between two mating surfaces (for both rolling and sliding friction) to reduce frictional resistance and wear. Further developments in technology and science have led to a number of solutions for reducing sliding friction resistance through effective lubrication with liquid (hydrostatic or hydrodynamic) or gas (gasostatic or gasodynamic). These technologies have been known and refined for decades and are used in both current and advanced machine and equipment designs. However, in many cases, the design of the equipment, the materials used or the intended operating conditions make it impossible to use conventional lubrication techniques. In addition, the current trends in modern tribology clearly point to a maximum limitation or reduction of the proportion of liquid lubricants used, mainly for environmental reasons [1].

In order to facilitate a comprehensive understanding of the solid lubrication topic, it is imperative to briefly revert to the surface itself and to undertake a more detailed examination of the contact area. In the context of engineering surfaces, irrespective of the machining or modification method, the attainment of perfect 'flatness' is not feasible. In addition, these surfaces are distinguished by a certain degree of inhomogeneity with regard to structure, chemical composition and properties on the cross-section. These include roughness, the presence of gas molecules such as water vapour, hydrocarbons and oxygen, adsorbed on the surface, surface layers of metal oxides (a few to a hundred nanometres thick), a hardened layer of deformed material (sheared or strained) and the final non-deformed base material. As a consequence the real contact area between the elements of the friction contact is considerably smaller in comparison with the apparent contact area. Thus, it can be summarised that the

© The Author(s), under exclusive license to Springer Nature Switzerland AG 2025
D. Batory, *Tribology of Low Friction Carbon Based Coatings*, Engineering Materials,
https://doi.org/10.1007/978-3-031-95979-0_2

actual contact area is contingent on the surface topography, material properties and loading conditions of the friction pair. The prevailing knowledge in this field suggests that the key to achieving a low coefficient of friction is to minimise the actual contact area to the greatest extent possible and that the actual contact area is defined by the roughness asperities of the friction pair elements that are in contact. In the instance that average contact stresses are found to be elastic, it is possible for local stresses to exceed the elastic limit at the tips of asperities on the surface roughness. This may consequently result in local plastic deformation. Moreover, it is believed that adhesion plays a role in the plastic deformation of asperities. This issue was examined by Williamson and Greenwood, who developed the plasticity index (Ψ) to address it [2, 3]:

$$\Psi = \frac{E^*}{H}\sqrt{\frac{\sigma}{R}} \qquad (2.1)$$

where H is the hardness (Pa), σ is the combined RMS surface roughness (m), and R is the mean asperity radius of the curvature. E^* is the combined plane stress modulus for the two surfaces (Pa):

$$\frac{1}{E^*} = \frac{1-v_1^2}{E_1} + \frac{1-v_2^2}{E_2} \qquad (2.2)$$

where $E_{1,2}$ are the elastic modulus and $v_{1,2}$ are the Poisson's ratio of contacting materials 1 and 2, respectively. Following a series of experiments conducted on surfaces of varying topographies, the researchers ascertained that the value of plasticity index can vary between 0.25 and 30. It was determined that the surface would deform plastically for $\Psi > 1$ and elastically for $\Psi < 0.6$, whereas in the region $0.6 < \Psi < 1$ the contact type is uncertain.

The model proposed by Williamson and Greenwood also enables to estimate the real contact area (A_r) [2, 4]. For elastic contacts A_r can be calculated according to Eq. 2.3:

$$\frac{A_r}{A_a} \cong \frac{3.2 P_H}{E^*\sqrt{\frac{\sigma}{R}}} \qquad (2.3)$$

where P_H is the mean Hertzian contact pressure. In the case of plastic deformation one can assume that A_r is proportional to normal load, inversely proportional to hardness, and independent of apparent contact area (A_a). Therefore, it can be estimated based on Eq. 2.4:

$$\frac{A_r}{A_a} \cong \frac{P_H}{H} \qquad (2.4)$$

In general this, and other rough-surface contact models [3, 5, 6] made an important contribution to the general discussion on the possibilities of reducing the adhesive forces and in a consequence the overall friction by lowering A_r and increasing H and E. Nevertheless, the fundamentals of contact mechanics modelling of surfaces being in the sliding contact extend well beyond the primary focus of this book. It should be noted, however, that the introduction of frictional forces introduces a significant increase in the complexity of the model. The main difference between static (indentation) and dynamic (sliding) character of the contact stresses is that in the case of the latter the stress field ahead of the counterbody is compressive, whereas the one behind is tensile. The introduction of friction has been shown to increase the maximum subsurface shear stress, thereby moving its position closer to the surface. This, in turn, even more complicates the contact model, as the coating-substrate interphase has to be considered. Nevertheless, in order to undertake further exploration of solid lubricants, it is imperative to ascertain that the local plastic deformation of the substrate material does not exert a detrimental effect on the integrity of substrate-coating system [4]. Other phenomena related to the mechanical interaction of the counterbody with the substrate material have been deliberately omitted. These include e.g. the formation/removal of surface oxide layers, changes in microstructure caused by the friction force, formation of adhesive contacts and material transfer between the cooperating surfaces. These phenomena are often individual properties of the material and do not necessarily account for its poor lubrication performance.

Several hundred years ago, Amontons formulated the first law of friction, which states that for most materials, the value of the friction coefficient is independent of the load on the friction node. Nevertheless, presently it is commonly known that a number of materials, including polymers, diamond, ceramics and thin, solid lubricating films on hard substrates, exhibit load-dependent friction behaviour. In a significant number of cases, the friction coefficient has been observed to decrease with increasing load. In order to quantify the relationship between the friction coefficient and the load the Hertzian contact model can be used [7]. It relates the friction coefficient to the material (shear strength) and mechanical properties of the substrates in contact (contact pressure). According to the theory developed by Bowden and Tabor [8] the friction force (F_t) is a product of the real contact area (A_r) and the shear strength of the material (τ). Therefore, the coefficient of friction can be expressed as the ratio of friction force to the normal load (W), or the ratio of shear strength to contact pressure, if both forces are divided by the contact area:

$$\mu = \frac{F_t}{W} = \frac{A_r \cdot \tau}{W} = \frac{\tau}{P_H} \tag{2.5}$$

The shear strength of solids at high pressures has been observed to have a pressure dependence given by the approximation:

$$\tau = \tau_0 + \alpha P_H \tag{2.6}$$

where τ_0 is the interfacial shear strength, a velocity accommodation parameter (an individual property of the interface), and α represents the pressure dependence of the shear strength [4]. Thus, the final expression for the coefficient of friction is as follows:

$$\mu = \frac{\tau_0}{P_H} + \alpha \tag{2.7}$$

The contact pressure, for a *sphere-on-flat* elastic contact (typical of bearing design) can be calculated as the mean Hertzian pressure and Eq. 2.7 can be rewritten as:

$$\mu = \tau_0 \pi \left(\frac{3R}{4E^*}\right)^{\frac{2}{3}} W^{-1/3} + \alpha \tag{2.8}$$

where R is the sphere radius. The analysis of Bowden and Tabor on Hertzian contacts provides clear evidence that the coefficient of friction is proportional to the load of friction contact. This finding is in direct opposition to the initial law of friction proposed by Amonton. For contacts loaded below the elastic limit the friction coefficient decreases with the increasing load (or mean Hertz pressure):

$$\mu \propto W^{-1/3} \tag{2.9}$$

The validity of this relationship has been corroborated through experimental methods by Bowden and Tabor (for a thin indium film applied to a hard steel substrate) and by other researchers working on solid lubricant coatings tested under macroscopic normal load range [4, 7, 8]. Namely, despite the elevated load on the friction node, the deformation of the elastically stiff and hard substrate material leads to a marginal increase in the contact area. Consequently, the value of coefficient of friction decreases.

In summary, according to the Bowden and Tabor concept, the ideal scenario for achieving low friction is as follows: an elastically stiff and hard substrate should be utilised in order to support the normal load and minimise the contact area. The surface coating must provide shear accommodation and reduce adhesive junction strength.

2.1 Solid Lubricant Coatings

The concept of an ideal friction pair, as presented in the previous chapter, is essential from the point of view mechanical interaction between the surfaces that are in contact. However, it does not consider the complete set of mechanical, physical and chemical properties of the coatings, nor their interaction with the surrounding environment. These are individual characteristics, and it is often these that account for their high application potential as solid lubricants. One of the fundamental mechanisms that occur in friction contacts based on solid lubricants is the formation of a transition layer (so-called tribofilm) transferred to the counterbody and covering the entire contact area. This results in a complete separation of the mating surfaces and also

2.1 Solid Lubricant Coatings

provides a low shear at or near the sliding interface, as long as the transfer film is not worn out or replenished. Furthermore, friction induced changes in coating structure and surface interaction with the atmosphere, known as tribochemical reactions can also lead to a further reduction in friction force. However, as previously mentioned, these are individual properties of those materials, and this chapter will focus on them.

Solid lubricant coatings are widely used to control friction and wear, especially in difficult application conditions such as high speeds and loads on mating parts, a wide temperature range or a corrosive working environment. In many cases, weight or price considerations, as well as service and lubricant replenishment options, lead designers to implement solid lubricant solutions. This is the case, for example, in space applications where lubrication requirements in vacuum or extreme temperature conditions can only be met by the use of solid lubricants [1, 9, 10].

Despite the broad spectrum of solid lubricant materials, they can be categorised within two groups. Namely soft and hard. Soft coating materials include polymers, soft metals, TMD (Transition Metal Dichalcogenide) compounds in the form of coatings or 2D materials (graphite/graphene, MoS_2, WS_2). Hard coating materials include mainly diamond-like carbon coatings or certain oxides. The synthesis technologies of many of these materials have been optimised to provide better lubricating properties under specific operating conditions. Most of these have been extensively reported in the world literature in recent years [1, 4, 9, 11–14]. Figure 2.1 provides a chronological map of the development of low-friction materials from a scale perspective.

Due to the nature of the book handed over to the reader, issues concerning carbon-based low-friction materials will be omitted from this chapter. It is intended to give the reader a general but not superficial introduction to the phenomenon

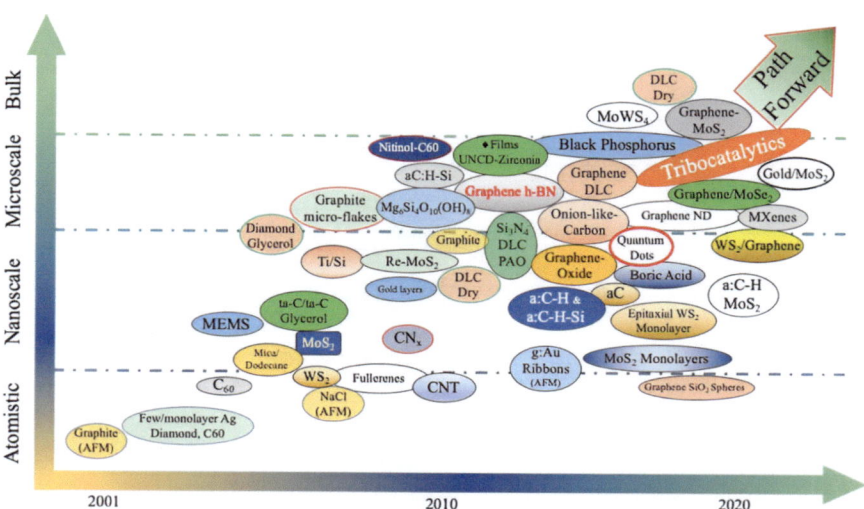

Fig. 2.1 Chronological evolution of superlubricity over length scales. Reprinted with permission under Creative Commons Attribution CC BY from [13]

of the mechanisms underlying the tribological properties of coating lubricants that improve the performance, durability and environmental compatibility of many moving mechanical systems.

Undoubtedly, the dynamics of the development of low-friction coating materials in terms of their properties and applications have been dominated by technological limitations in the possibilities of their synthesis and optimisation. The development of surface engineering methods has opened up new possibilities for shaping the tribological properties of machine and equipment assemblies. A wide range of technologies for the chemical and physical deposition of coatings from the gas phase, as well as increasingly advanced research methods in solid state physics, have led to many breakthroughs in this area. In addition to new coating materials, significant progress has been made in optimising the chemical and phase composition of nanostructured, multi-phase, intelligent, adaptive coatings with differentiated architectures. New variations in synthesis technology have made it possible to apply technical coatings to the surfaces of components made of heat-sensitive materials. Figure 2.2 gives a historical overview of the development of tribological coatings and low-friction films.

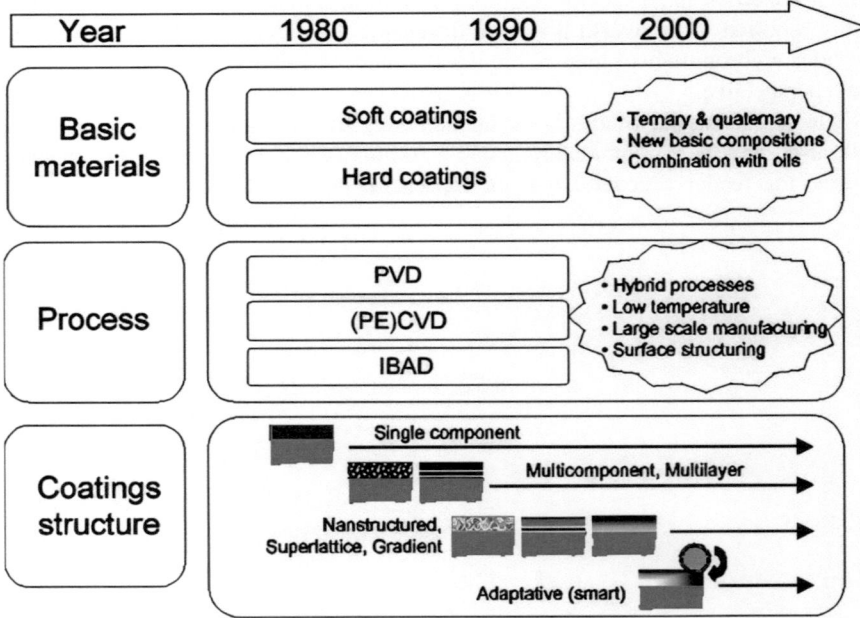

PVD = Physical vapor deposition
(PE) CVD = (Plasma enhanced) chemical vapor deposition
IBAD = Ion-beam-assisted deposition

Fig. 2.2 Historical development of tribological coatings and solid lubricant films. Reprinted with permission from [1]

2.1.1 Polymer Coatings

Typically, polymeric materials have low mechanical properties compared to most other coating materials. This is even more true in tribological applications where friction and wear are critical. The significantly lower ratio of hardness to Young's modulus of polymeric materials means that whatever the nature of the friction pair (polymer-metal, polymer–polymer), the contact between them is usually elastic and the wear resistance is significantly lower. The fact that the mechanical properties of polymers often change over time is also of considerable importance. On the other hand, the excellent chemical resistance of these materials or their relative resistance to high temperatures should not be overlooked. For these and other reasons, polyethylene (high density HDPE and low density LDPE) and polytetrafluoroethylene (PTFE) are the most widely used base materials for solid lubricants and bearing components [15–17]. The coefficient of friction of these polymers in contact with themselves or metals is usually in the range of 0.1–0.5, but there are literature reports of values outside this range. Slight variations in sliding speed, temperature, or normal load on the friction node have been shown to have a significant impact on this parameter, with the potential to result in substantially different values. This variability can also be influenced by the surface roughness of the friction pairing elements. The phenomenon of low friction coefficient values can be attributed to the linear and unbranched molecular structure of the substance, as well as its weak intermolecular bonds. In the friction process, the adhesive component of the coefficient of friction is of great importance, leading to the formation of a thin polymeric transition layer on the surface of the counterbody, which significantly reduces the degree of wear [4]. The intermolecular bonds of the polymer break and form active groups, which react chemically with the mating surface and deposit on it. Further contact between the friction pair results in an increase in the thickness of the transition layer. This is due to the cohesion forces at the interface often being stronger than those in the polymer coating itself. In addition, the molecular chains in the growing crystalline structure are arranged parallel to the sliding direction. This results in a constant and low coefficient of friction [18, 19]. PTFE coatings have a low coefficient of friction in both vacuum and air atmospheres. However, due to their low surface free energy and low thermal conductivity, these coatings are susceptible to damage in the form of melting, which limits their use to low speed applications. Improving the wear resistance of PTFE coatings without risking an increase in the coefficient of friction values can be achieved by, among other things, doping the PTFE coating matrix with different types of fillers [20, 21], or adding PTFE in the form of fibres or particles as a solid lubricant to the coating matrix of composite materials [21–23]. Figure 2.3 shows an example of the surface morphology and cross section of a Ni-PTFE coating.

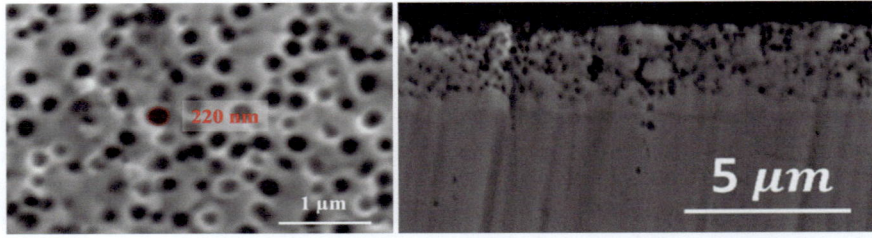

Fig. 2.3 Surface morphology and cross section of a Ni-PTFE coating. Reprinted with permission under Creative Commons Attribution CC BY from [23]—combined Figs. 2.2b and 2.4d

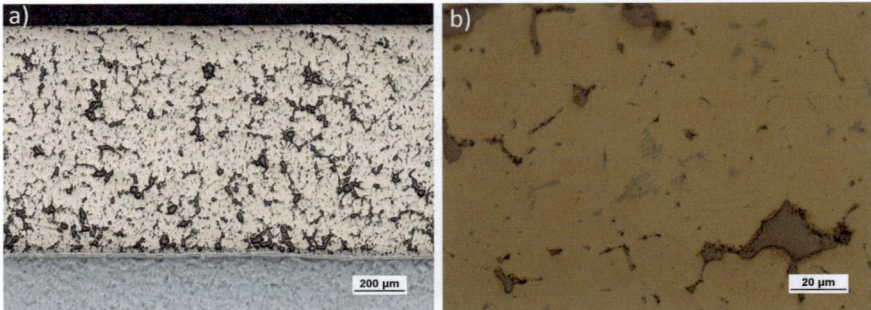

Fig. 2.4 **a** $CuPb_{24}Sn$ solid lubricant coating on steel substrate; **b** microstructure of low-melting eutectics

2.1.2 Soft Metal Coatings

The application of low-friction coatings derived from soft metals to high-hardness surfaces enables friction nodes to function with a coefficient of friction of approximately $\mu \sim 0.1$. Lead, tin, antimony or indium are widely used as alloying elements in babbitt, lead bronze or Al–Sn coatings [24–28]. The main feature that makes these metals suitable for use as solid lubricants is their relatively low shear strength, which decreases with increasing temperature. The elevated temperature at the point of frictional contact also signifies that the low-melting eutectics of bearing alloys can undergo localised overmelting at the contact level, which is characterised by surface roughness peaks. This phenomenon can contribute to a further reduction in frictional resistance. The abundance of easy-slip systems embedded within their crystallographic structure, in conjunction with their low melting point, which fosters frictional recrystallisation processes under heat, indisputably contributes to their remarkable efficiency as solid lubricants. Figure 2.4 presents a microstructure of a heavy-duty self-lubricating bimetal bearing with a high-load capacity bronze alloy sintered onto a steel backing. From the application point of view, the operating temperature is important when selecting solid lubricants. This is because, in the case of some soft metals (tin or indium), its beneficial effect in reducing shear strength is offset by progressive degradation through oxidation processes. Elements such as Ag, Au or Pt perform much better in this respect. Due to their high chemical and thermal inertness, these materials have a wide application potential. The use of solid lubricants based on soft metals appears to be more suitable for applications in friction nodes operating under rolling friction and high temperatures. Coatings based on silver, barium or gold are successfully used on the surfaces of low-load bearing components operating in x-ray tubes or friction assemblies operating in space conditions [29].

2.1.3 Lamellar Solids

Low friction coatings that have achieved commercial success are undoubtedly MoS_2 coatings synthesized by means of physical vapour deposition methods. Like graphite, MoS_2 or WS_2 coatings owe their excellent lubricating properties to their hexagonal crystallographic structure. Characteristic of this lattice is the lamellar arrangement of densely packed atomic planes, shown in Fig. 2.5, parallel to the surface of the elementary cell base.

This highly anisotropic crystalline structure consists of a layer of molybdenum atoms arranged in a hexagonal pattern, with each molybdenum atom surrounded at equal distances by six sulfur atoms located at the corners of the tetrahedron. Strong covalent bonds are present between the S–Mo–S planes, while slippage can only occur between individual sequences of MoS layers, which are only connected by weak van der Waals forces. According to the theory of slip-induced plastic deformation of crystalline materials, under the action of tangential forces, the individual base planes spread out one after the other on the surface of the counterbody [4, 30]. Deformation by sliding in the other crystallographic directions, including the direction perpendicular to the base, is virtually impossible. The focus of extensive research and industrial interest in these materials over many decades is primarily attributable to their exceptional lubricating properties when subjected to conditions of low pressure. This quality renders them highly suitable for deployment in the context of friction management in space-related applications [10]. The coefficient of friction of MoS_2 coatings observed in ultra-high vacuum or in a controlled gas atmosphere is

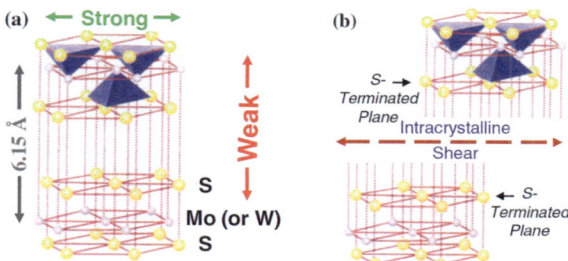

Fig. 2.5 a Lamellar crystallographic structure of molybdenum (or tungsten) disulphide, b shear mechanism and degradation of weak van der Waals bonds between the crystallographic planes. Reprinted with permission from [4]

less than 0.05, while the wear rate is relatively low. However, the operating environment, particularly humidity, exerts a substantial influence, resulting in a considerable deterioration of their tribological properties, as evidenced by a decline of both friction coefficient and wear resistance. Reactions of unsaturated chemical bonds at the edges of the base planes with atmospheric oxygen can lead to oxidation and the formation of MoO_3 compounds [1, 4, 11]. The technologies utilised in the synthesis of MoS_2 coatings permit the extensive manipulation of their structure, chemical and phase composition. As a result, the coatings are not always characterised by a fully crystalline structure and an appropriate crystallographic orientation with respect to the slip direction. Nevertheless, the temperatures and stresses generated at frictional contact lead to energetically favoured tribochemical and tribomechanical reactions, resulting in crystallisation of the amorphous phase of the coating or reorientation of pre-existing crystallites along the slip direction, as shown in Fig. 2.6.

A promising strategy to reduce the susceptibility of MoS_2 coatings to environmental degradation is to dope them with metallic materials or oxides.

A substantial body of literature attests to the enhancement of properties of nanocomposite coatings of the MoS_2/Me (Me: Ti, Cr, Au) or MoS_2/Sb_2O_3/Au type, in the context of high humidity conditions. This enhancement is characterised by an increase in density, hardness and oxidation resistance [4, 31–33]. It can thus be concluded that the fundamental mechanisms of the low friction properties of molybdenum disulphide-based coating materials are based on the formation and subsequent reorientation of crystallographic planes (0002) in a direction parallel to the contact plane. These planes are deposited on the mating material as a result of shear, forming a transition layer that facilitates mutual sliding.

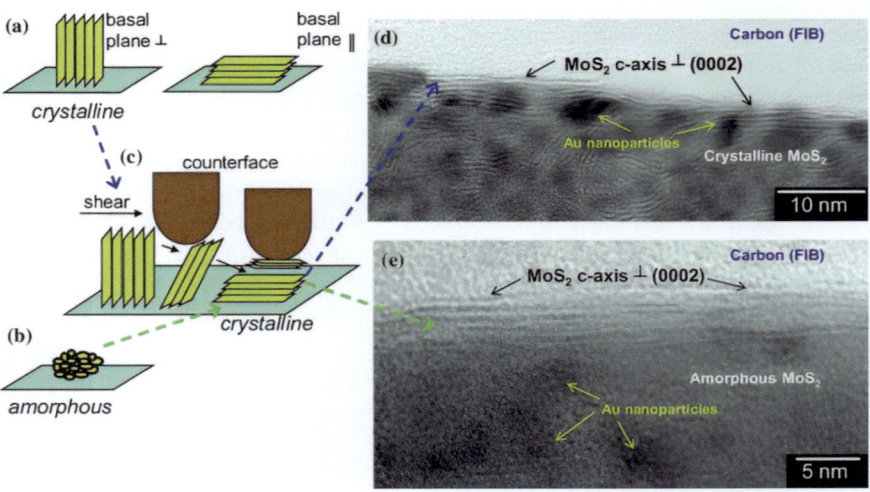

Fig. 2.6 Schematic representation of the crystallisation process of the amorphous structure of the coating and the reorientation of the crystallites formed according to the direction of friction pair slip. Reprinted with permission from [4]

2.1.4 Self-healing Coatings

The protective coatings applied to engineering materials are susceptible to damage accumulation during their service life, despite the protective mechanisms they provide. Consequently, under extreme conditions such as fatigue, wear or an aggressive working environment, they may fail once damage exceeds a critical value. The progression of mechanical wear and tear that occurs during standard operation is indicative of the overall service life of a coating. Nevertheless, a multitude of additional factors, including microcracks or other damage arising from defects in the microstructure of the coating, short-term absence of lubrication or the presence of abrasive particles at the contact area, have the capacity to substantially diminish the reliability and service life of the tribological system. In order to address the aforementioned issues, a new generation of coating materials is required that exhibits self-healing properties or possesses the capacity to reverse the accumulation of damage. The implementation of such materials would result in a significant enhancement of sustainability, accompanied by substantial savings in resources and energy. The self-healing process is contingent on the capacity of the materials to undergo autonomous repair. This capacity is typically prompted by the occurrence of damage itself or by external stimuli, such as load, heat, working atmosphere or their simultaneous combination. In this respect transition-metal dichalcogenides (TMD), particularly WS_2 coatings have a high application potential. Their excellent tribological properties, like those of MoS_2 coatings, are due to their unique anisotropic crystalline structure. They are also known for their ability to reorient atomic planes under frictional contact forces in a direction parallel to the coating surface. As demonstrated in Fig. 2.7, the wear products of WS_2-based coating may accumulate in an artificially created notch in the coating surface. Subsequent reorientation and bridging of these products may then result in the formation of a continuous tribofilm, which, in turn, enables damage to be repaired. The WS_2 platelets exhibit a high degree of adaptability, with the capacity to flexibly overspread the curved coating/notch interface. Furthermore, at the top of the healed notch, they demonstrate a marked reorientation, aligning themselves parallel to the sliding direction [34, 35].

Functional ceramic coatings are susceptible to cracking when subjected to external mechanical stresses, a condition that can be further exacerbated by thermal stresses. The initiation of a self-healing process has emerged as a promising approach for mitigating failure in ceramics experiencing mechano-induced crack propagation. Moreover, surface self-organisation of lubricious ternary oxide based films during sliding contact appears to possess the potential to generate both a self-healing and a self-lubricating properties. Nevertheless, there exists a series of prerequisites that must be fulfilled in order to successfully create the self-healing coating: (i) the presence of a mobile phase capable of migrating to the defect site to perform the required material reconstruction; (ii) as a consequence of the migration of the mobile phase, which is stimulated by thermal and/or mechanical means, a ternary oxide layer should be formed; (iii) the newly formed oxide should exhibit low shear resistance when subjected to sliding friction conditions [36]. It has been demonstrated that the

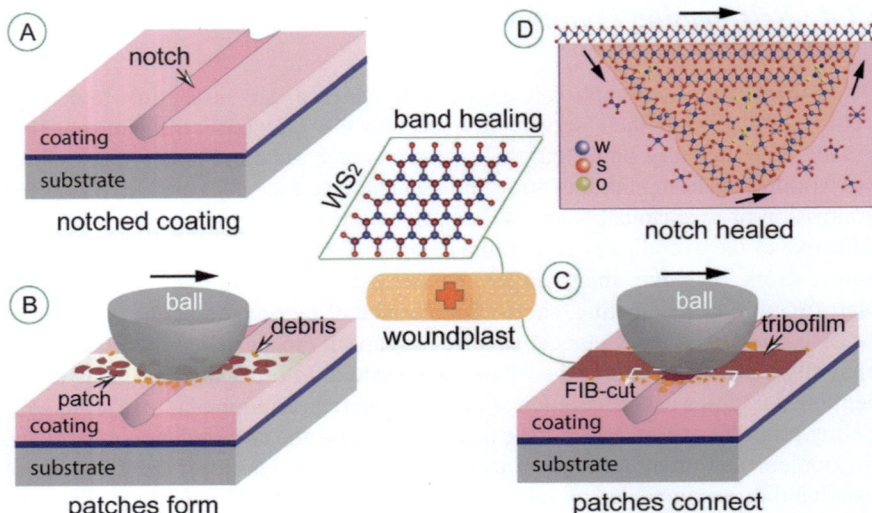

Fig. 2.7 Schematic of the self-healing process of a nanocomposite coating: **a** a pre-notch made in WS2/a-C coating; **b** patchy tribofilms start to form; **c** tribofilms interconnect and fill into the notch; **d** the cross-section image illustrating the notch, which has been filled with WS$_2$ platelets—the platelets have adapted to the curved shape of the notch and are aligned with the sliding direction. Reprinted with permission from [34]

self-healing properties of niobium oxide (Nb_2O_5) are activated in the presence of silver through the formation of Nb–Ag–O ternary oxide. The surface reconstruction process is initiated in the wear track at a significantly lower temperature than would be required from a thermodynamic perspective. This observation indicates that the self-healing process was initiated by mechanically induced stress, and that this may represent a novel approach to enhance the wear and crack resistance characteristics of ceramic components. Moreover, it is suggested that this process can be calibrated to yield the desired frictional response, thus opening up new potential applications of these materials in space, aerospace, power generation, and defence [37].

Recently, catalytic nanocomposite MeN–Cu coatings were presented as a means to facilitate the in-situ formation of protective films, structurally similar to diamond-like carbon (DLC), directly at the sliding interfaces. The formation of carbon-based tribofilm was achieved via the dissociative extraction from base-oil molecules on a catalytically active MoN_x–Cu nanocrystalline coating. A tribological examination of the resultant tribofilms revealed that, in comparison with zinc dialkylditiophophate, they exhibit a significantly reduced wear rate and lower friction. Furthermore, molecular dynamics simulations have confirmed that the protective layer is formed due to the presence of catalytically-reactive copper clusters exposed to hydrocarbon sources at the contact interfaces during sliding [38]. The next set of test results involved VN–Cu, MoN–Cu and MoVN–Cu coatings operating in an alkane environment. This is particularly interesting from the point of view of the relevance of

2.1 Solid Lubricant Coatings

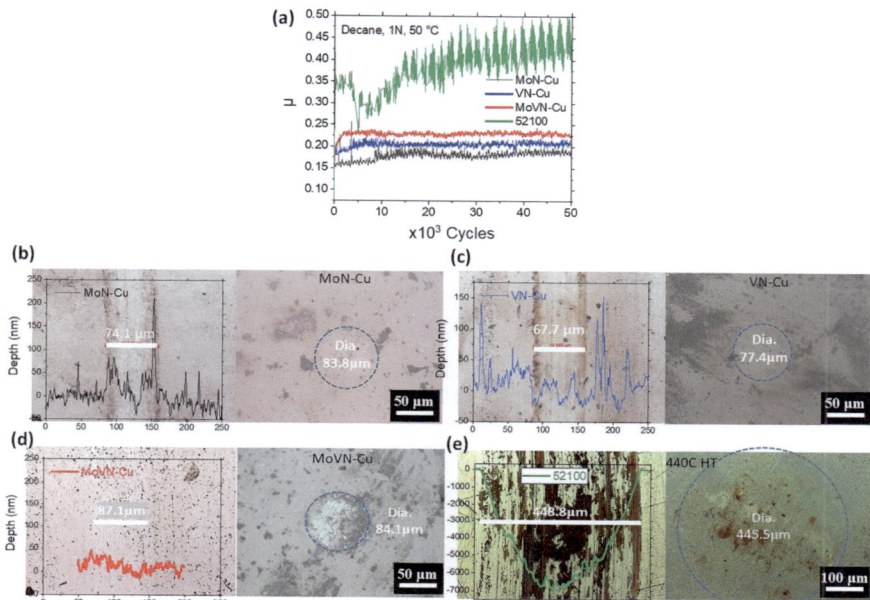

Fig. 2.8 **a** Coefficient of friction values for films tested in decane at 1 N load and 50 °C. Optical micrographs of the contact areas and inserted stylus profilometry results of the wear tracks formed on **b** MoN–Cu, **c** VN–Cu, **d** MoVN–Cu films, and **e** 52,100 steel substrate used as the reference sample. Reprinted with permission under Creative Commons CC BY without any changes from [39]

tribocatalysis to fuel-based lubrication of sliding components in internal combustion engines. Figure 2.8 shows a comparison of the coefficient of friction and optical microscope images of wear tracks and wear scars, recorded after the tests in the decane environment.

2.1.5 Self-organized, Adaptive Chameleon Coatings

As previously discussed in this text, the low shear strength—a defining characteristic of nearly all solid lubricants, including lamellar solids and further discussed carbon coatings—occurs only in a very limited range of ambient conditions. This may, in turn, represent a significant limitation, as their favourable tribological performance is contingent upon the specific test environment and operational conditions. Molybdenum disulfide coatings exhibit the lowest friction coefficient under vacuum and dry atmospheres. However, an increase in humidity noticeably deteriorates their tribological performance. Conversely, graphite exhibits excellent lubricating properties in the presence of moisture or other condensable vapours. However, it is not well tolerated in dry or vacuum atmospheres, which leads to an increase in friction and wear. A common factor affecting the tribological performance of these two and many other solid lubricants is thermal degradation [40, 41]. The overarching concept of surface engineering, with regard to the creation of wear-resistant and anti-friction layers, entails the development of coatings that are specifically designed for particular tribological systems. These coatings are engineered to withstand significant external impact while retaining their original properties, or exhibiting only minimal changes, throughout their entire period of use [42]. It is important to note that in many applications, the operating conditions can change drastically. In such cases, the coatings may be required to provide low friction and wear over a wide range of humidity, vacuum conditions, or extremely high loads and operating temperatures. For instance, coatings designed for aerospace applications are typically intended to function effectively in a vacuum environment. However, during the launch of a satellite, these coatings may be exposed to moisture and significant temperature fluctuations [43]. The most prevalent cause of accelerated tool deterioration is the utilisation of contemporary high-performance machining and materials forming techniques. These techniques frequently entail the operation of tools at temperatures and under stresses that deviate significantly from equilibrium [44]. The heightened necessity for lightweight, high-strength materials in energy-efficient vehicles has prompted numerous recent investigations into the enhancement of friction and wear resistance in aluminium, titanium, and magnesium-based alloys, which are typically regarded as exhibiting suboptimal tribological performance, particularly under varying high-temperature operational conditions [45].

In a multitude of scenarios, conventional lubricants, such as oils and greases, become ineffective due to evaporation or chemical degradation when subjected to the aforementioned conditions. Traditional solid lubricant coatings often prove inadequate in providing the requisite protection and performance, given the potential for variable and significantly disparate working conditions. Consequently, there is a clear need for coatings that can adapt and exhibit self-protection and self-healing properties, thereby ensuring optimal friction and wear performance across diverse environmental settings.

The indisputable factors in this regard, which extend far beyond the commonly accepted mechanical response of the system to given working conditions, are extensive physical and chemical interactions between frictional bodies as well as with the surrounding atmosphere [46].

This leads us to a new generation of chameleon coatings, or tribo-systems, which exhibit adaptive behaviour. Typically, these are nanocomposite hard coatings with embedded solid lubricant reservoirs that automatically and reversibly respond to external mechanical, thermal and physical forces by adjusting their surface composition and morphology, thereby improving the wear and friction characteristics of a tribo-couple under given working conditions. Figure 2.9 illustrates the reversible self-adaptive process of surface chemistry and structure in a chameleon coating in response to environmental changes. The WC/DLC/WS$_2$ composite coating demonstrated high resistance to wear in both vacuum and air environments, as well as excellent friction recovery in humid and dry environmental cycling [47]. The plasma

2.1 Solid Lubricant Coatings

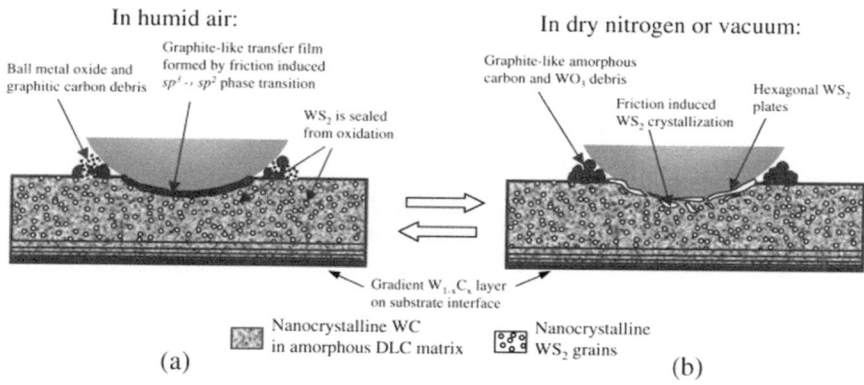

Fig. 2.9 Mechanisms of a reversible self-adaptation in surface chemistry and structure of WC/DLC/WS$_2$ nanocomposite coatings with the environment change from humid air (**a**) to dry nitrogen or vacuum (**b**). Reprinted with permission from [47]

electrolytic oxidation (PEO) process of aluminium alloy has recently been found to provide high hardness and load support for burnished graphite-MoS$_2$-Sb$_2$O$_3$ chameleon solid lubricant coating, while simultaneously protecting it against high temperature oxidation. As a consequence of thermo-mechanical interactions occurring at the friction interface, the solid lubricant effectively filled the voids present in the porous structure of the PEO coating, thereby forming a protective transfer layer between the cooperating elements of the friction pair. Furthermore, the synergistic effect between the PEO coating and the chameleon layers promoted defect healing and adaptive behaviour of the coating [45].

In general the adaptation mechanisms include self-guided evolutions toward formation of low friction and wear contact conditions according to the following applied strategies [48]:

- temperature activated diffusion of metal lubricants to the surface
- temperature and environmentally activated formation of lubricious oxide phases
- temperature and strain actuated structural evolutions in the contact.

In many aspects these strategies may complement each other by facilitating the formation of oxides with simultaneous diffusion of noble metals as well as temperature and strain induced structural evolution of initially amorphous matrix [4, 32, 40, 41, 44].

Figure 2.10 illustrates a schematic of temperature-activated diffusion of gold, together with the temperature-induced structural evolution of MoS$_2$ from an amorphous MoS$_2$–Sb$_2$O$_3$ matrix.

This phenomenon is further exemplified by the impact of V and Ag diffusion, which resulted in a temperature-adaptive tribological behaviour of the VAlN/Ag multi-layer coating. The synergistic diffusion of V and Ag elements at elevated temperatures stimulated the in-situ formation of distinctive Al$_2$O$_3$/AgVO$_3$–Ag$_3$VO$_4$/

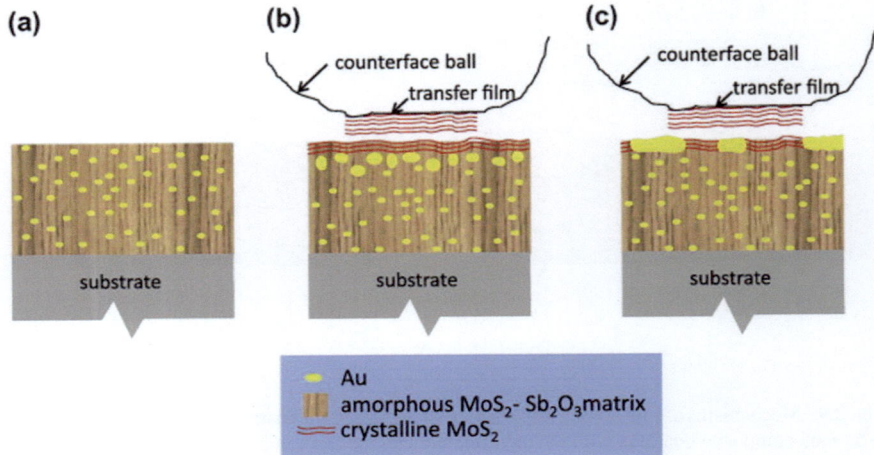

Fig. 2.10 A schematic of temperature activated diffusion of gold together with temperature-induced structural evolution of MoS$_2$ from an amorphous MoS$_2$–Sb$_2$O$_3$ matrix. Reprinted with permission form [32]

Ag lubricant phases, which ensured lower CoF and high wear resistance. A wide temperature range tribological investigation revealed that the Ag upper layer, when in a softened state, exhibited a lubricating effect during low to mid-range temperatures, while the AgVO$_3$–Ag$_3$VO$_4$ compound provided lubrication during friction at 650 °C [49]. Figure 2.11 presents a schematic view of the evolution of element diffusion and the overall tribological mechanism behind it.

It is beyond the scope of this book to provide an exhaustive review of the existing literature on composite self-adapting or self-organising nanostructural materials, which have been the subject of numerous studies. It is evident that several of the solutions presented here, as well as other proposed solutions, are based on known low-friction materials. This constitutes the basis for further debate on the structure, morphology and mutual interaction of the coating components, with the aim of ensuring that the resulting tribological properties meet the criterion of self-adapting, chameleon or self-organising materials. The consensus among numerous literature reports on this topic is that future generations of hard protective coatings for extreme tribological applications will be complex adaptive systems capable of sustaining a multitude of external impacts and adapting to changing (and severe) operating conditions.

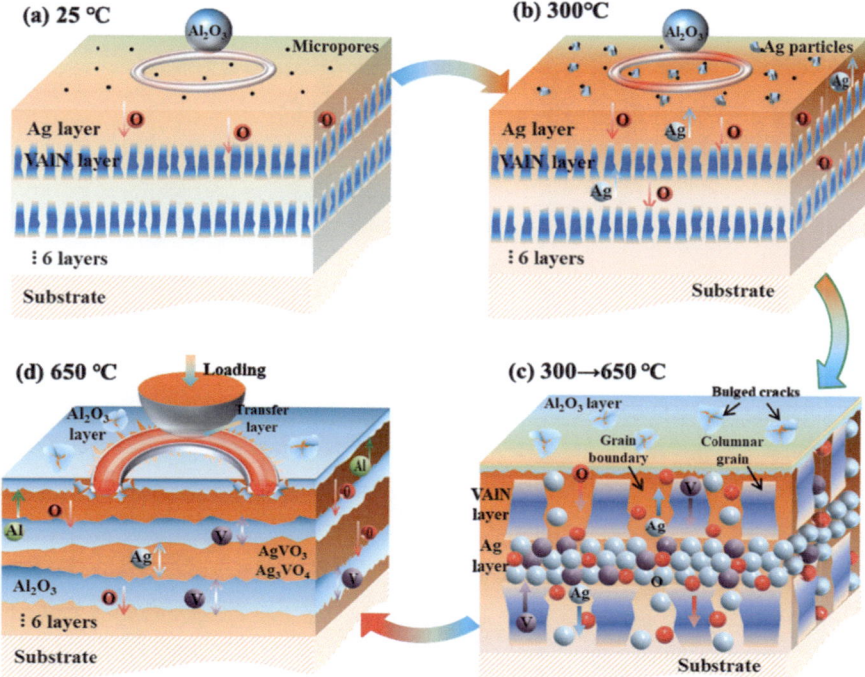

Fig. 2.11 The evolution of element diffusion and tribology mechanism of VAlN/Ag multi-layer coating over 25–650 °C. Reprinted with permission from [49]

2.2 Synergy of Solid Lubricants and Surface Texturing

Surface texturing is a technique employed to create the desired surface pattern, which ultimately determines the characteristics of the surface geometrical structure of the material. It is therefore possible that this may affect the friction and wear. The surface texture of a given element may be fabricated during the manufacturing process, for example through the use of injection molding, or subsequently modified through the addition of micro grooves, micro dimples, micro bumps, and microchannels of varying spacing, dimensions, shapes, width, distance, area fractions, and depth. In conditions of dry friction, surface textures have been demonstrated to have a number of effects. Firstly, they have been shown to reduce the real area of contact. Secondly, they have been demonstrated to trap wear debris. The combination of these effects leads to a reduction in friction and an extended wear life [50, 51]. Figure 2.12 presents images of different patterns of laser texturing on the stainless steel surfaces.

According to Rosenkranz et al. [52] the design of textures for dry friction conditions should take into account eight points:

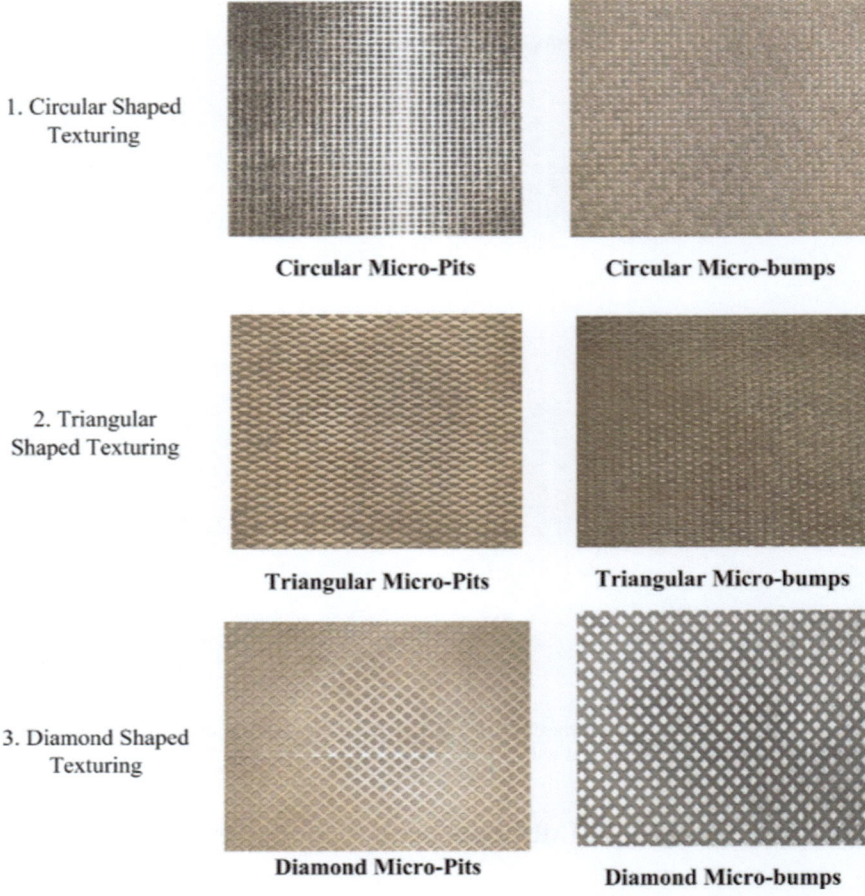

Fig. 2.12 Images of different patterns of laser texturing on the SS surfaces. Reprinted with permission form [51]

1. The distance between the textures should be sufficient to prevent the accumulation of debris and subsequent agglomeration. If not, the material could be ploughed into the cooperating surfaces.
2. It is essential that the coverage area ensures the presence of sufficient smooth areas to accommodate the normal load and maintain the load-bearing capacity.
3. It is inadvisable to create textures of excessive depth, as this may result in edge stresses that could have a detrimental impact on the material, potentially leading to excessive plastic deformation.
4. It is imperative that the dimensions of the textures correspond with those of the largest wear debris, as they must be accommodated.
5. It is imperative that the surface texturing technique generates the lowest possible stress levels.

2.2 Synergy of Solid Lubricants and Surface Texturing

6. It is important to understand the potential indirect effects of texturing on heat transfer, surface free energy and, consequently, friction and wear.
7. It is strongly advised that the surface texture design processes employ all available predictive tools, including the results of computer simulations, in order to accurately predict the influence of the applied surface modification on the resulting tribological properties.
8. While surface texturing can reduce the ploughing and adhesion components of the friction force under dry sliding conditions, further significant progress in reducing the friction force of textured surfaces can be achieved by using appropriate low-friction coatings.

In the domain of low-friction coatings, the final point signifies a prospective avenue for further exploration, providing a conduit for the implementation of well-established surface treatment methodologies within a wholly novel context. Although this approach may initially appear to be a nascent field with numerous unresolved inquiries, it nevertheless offers a promising opportunity for the advancement of wear and friction reduction in sliding contacts.

Two conceptual approaches are employed when combining textures with solid lubricant coatings. The initial approach assumes that the coating itself may be textured, either directly during the deposition process via the shielding effect of special, typically metal meshes with known grid intervals (Fig. 2.13) [53], or by subsequent treatment by laser texturing (Fig. 2.14) [54, 55], chemical etching and litography [51], as well as micro-milling [56]. In the majority of cases, the authors have observed a reduction in the wear rate of the textured coating, which can be attributed to the smaller contact area and the trapping mechanism of wear debris. This results in the removal of debris from the tribological interface, thereby extending the service lifetime. This approach is typically employed for hard and wear-resistant coatings.

The second approach is predicated on the assumption that coatings can be deposited directly on textured substrates. Well-defined surface textures produced by lithography and selective etching, with subsequent deposition of hard and wear-resistant TiN and DLC coatings, have been reported. The positive and negative aspects of this approach were discussed, and it was concluded that under certain conditions, the addition of a texture to a sliding surface may be highly beneficial. However, under other conditions, the same texture may impair the contact situation, for example by altering the formation of the transfer layer, which under non-textured circumstances decreases the wear and protects the counterbody material [57].

The combined benefits of surface texturing, coatings, and solid lubricant incorporation were obtained for MoS_2 [58] and MoS_2–Ag [59] layers integrated into laser-textured micro-dimples. In both cases, the formation of a lubrication film, secondary lubrication through the supply of solid lubricant from the dimples to the surface, and the capture of debris were identified as the primary underlying mechanisms. Figure 2.15 presents the comparison of friction and wear mechanisms of the CrN–MoS_2–Ag coating for flat and textured surfaces.

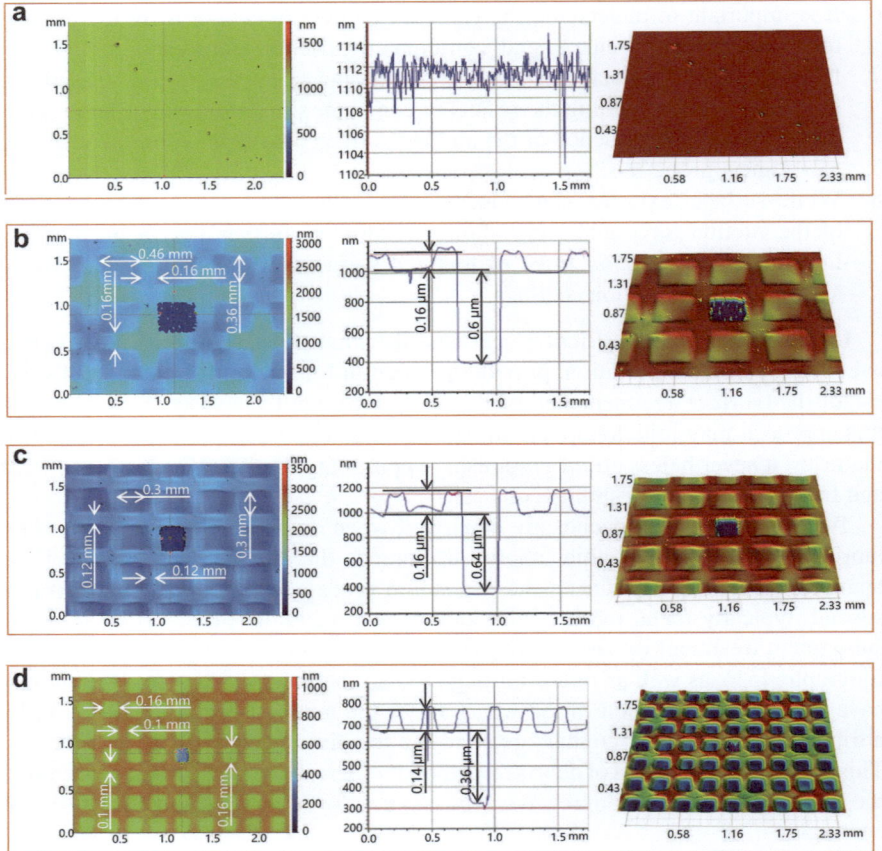

Fig. 2.13 Texturing by masking. Reprinted with permission from [53]

A further synergistic effect was observed in the case of a silver coating synthesised on the textured surface of a tantalum interlayer. It was found that the role of texture was not a significant factor in the reduction of friction at room temperature. However, at elevated temperatures, the silver lubricating phase in the wear track was gradually lost with progressive wear, while the textured dimples served as a reservoir for lubricant, ensuring a low and stable coefficient of friction [60].

Irrespective of the method of surface texturing and the resulting surface geometrical structure, these are invariably subject to progressive degradation and wear under the prevailing operational working conditions. Once the texture features have been worn down, the enhancement of the tribological properties becomes less and less visible. Therefore, maintaining the texture features is of critical importance for the durability and reliability of engineering components. Thus, the last aspect of synergy between surface texturing, coatings and solid lubricants is preserving the surface

2.2 Synergy of Solid Lubricants and Surface Texturing

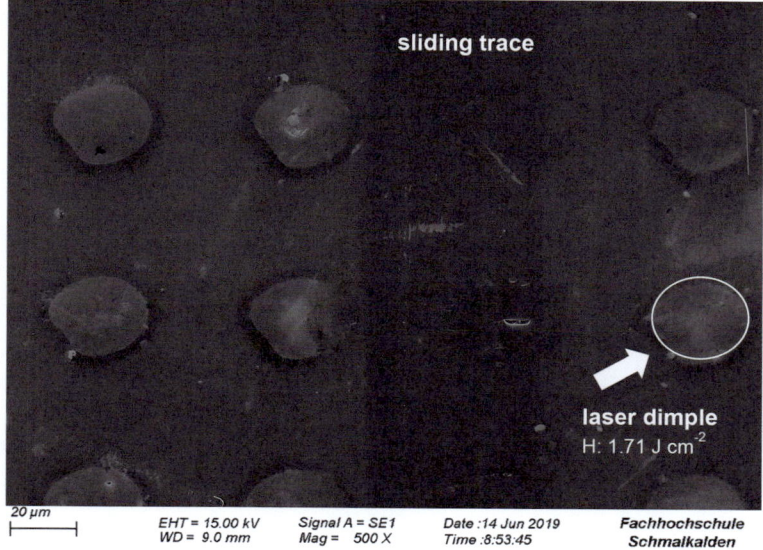

Fig. 2.14 Sliding trace in fs-laser structured a-C:H. Reprinted with permission from [54]

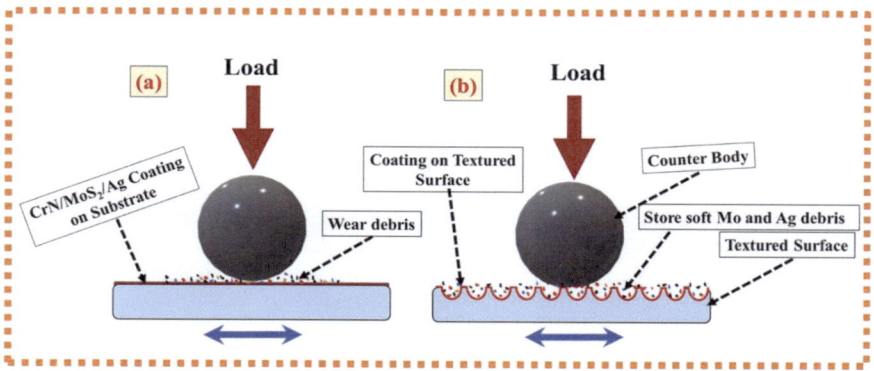

Fig. 2.15 Schematic illustration of friction and wear mechanism for CrN–MoS$_2$–Ag coating (**a**) and coatings on a textured surface (**b**). Reprinted with permission from [59]

texture, which can be ensured by a hard and wear-resistant coating that also functions as a solid lubricant.

The ongoing debate on the superiority of textured coating or coated texture in enhancing the friction and wear resistance of engineering surfaces was illuminated by Bondarev et al. [61]. The authors conducted a comparative analysis of the two approaches. Specifically, the texturing of a steel substrate with a femtosecond laser was conducted prior to the deposition of an a-C:H/WC coating, as well as the texturing of an a-C:H/WC coating that had already been deposited. Despite the absence of a statistically significant difference in tribological performance between the two approaches, irrespective of pattern type, temperature and frequency, the authors nevertheless indicated the second approach as more beneficial due to its technological and economic advantages for industrial applications when several types of substrates requiring different laser texturing parameters are utilised.

References

1. Donnet, C., Erdemir, A.: Solid lubricant coatings: recent developments and future trends. Tribol. Lett. **17**, 389–397 (2004). https://doi.org/10.1023/B:TRIL.0000044487.32514.1d
2. Greenwood, J.A., Williamson, J.B.P: Contact of nominally flat surfaces. Proc. R. Soc. Lond. A Math. Phys. Sci. **295**, 300–319 (1966). https://doi.org/10.1098/rspa.1966.0242
3. Taylor, R.I.: Rough surface contact modelling—a review (2022)
4. Scharf, T.W., Prasad, S.V.: Solid lubricants: a review. J. Mater. Sci. **48**, 511–531 (2013). https://doi.org/10.1007/s10853-012-7038-2
5. Greenwood, J.A., Wu, J.J.: Surface Roughness and Contact: An Apology (2001)
6. Whitehouse, D.J., Archard, J.F.: The properties of random surfaces of significance in their contact. Proc. R. Soc. Lond. A Math. Phys. Sci. **316**, 97–121 (1970). https://doi.org/10.1098/rspa.1970.0068
7. Singer, I.L., Bolster, R.N., Wegand, J., Fayeulle, S., Stupp, B.C.: Hertzian stress contribution to low friction behavior of thin MoS_2 coatings. Appl. Phys. Lett. **57**, 995–997 (1990). https://doi.org/10.1063/1.104276
8. Bowden, F.P.: Tabor D: Friction and Lubrication of Solids. The Clarendon Press, Oxford (1964)
9. Donnet, C., Erdemir, A.: Historical developments and new trends in tribological and solid lubricant coatings. Surf. Coat. Technol. **180–181**, 76–84 (2004). https://doi.org/10.1016/j.surfcoat.2003.10.022
10. Roberts, E.W.: Space tribology: Its role in spacecraft mechanisms (2012)
11. Donnet, C., Martin, J.M., Le Mogne, T., Belin, M.: Super-low friction of MoS, coatings in various environments (1996)
12. Zabinski, J.S., Sanders, J.H., Nainaparampil, J., Prasad, S. V: Lubrication using a microstructurally engineered oxide: performance and mechanisms. (2000)
13. Ayyagari, A., Alam, K.I., Berman, D., Erdemir, A.: Progress in superlubricity across different media and material systems—a review (2022)
14. Berman, D., Erdemir, A., Sumant, A.V.: Graphene: A new emerging lubricant (2014)
15. Guermazi, N., Elleuch, K., Ayedi, H.F., Zahouani, H., Kapsa, Ph.: Susceptibility to scratch damage of high density polyethylene coating. Mater. Sci. Eng. A **492**, 400–406 (2008). https://doi.org/10.1016/j.msea.2008.05.035
16. Ian, H., Philip, S.: Tribology, Friction and Wear of Engineering Materials (2017)
17. Song, W., Wang, S., Lu, Y., Zhang, X., Xia, Z.: Friction behavior of PTFE-coated Si_3N_4/TiC ceramics fabricated by spray technique under dry friction. Ceram Int. **47**, 7487–7496 (2021). https://doi.org/10.1016/j.ceramint.2020.10.255
18. Biswas, S.K., Vijayan, K.: Friction and wear of PTFE—a review. Wear **158**, 193–211 (1992). https://doi.org/10.1016/0043-1648(92)90039-B
19. Liu, X.-X., Li, T.-S., Liu, X.-J., Lv, R.-G., Cong, P.-H.: An investigation on the friction of oriented polytetrafluoroethylene (PTFE). Wear **262**, 1414–1418 (2007). https://doi.org/10.1016/j.wear.2007.01.021
20. Tang, A., Wang, M., Huang, W., Wang, X.: Composition design of Ni-nano-Al_2O_3-PTFE coatings and their tribological characteristics. Surf. Coat. Technol. **282**, 121–128 (2015). https://doi.org/10.1016/j.surfcoat.2015.10.034
21. Friedrich, K.: Polymer composites for tribological applications (2018)

References

22. Balaji, R., Pushpavanam, M., Kumar, K.Y., Subramanian, K.: Electrodeposition of bronze-PTFE composite coatings and study on their tribological characteristics. Surf. Coat. Technol. **201**, 3205–3211 (2006). https://doi.org/10.1016/j.surfcoat.2006.06.039
23. Lee, M., Park, J., Son, K., Kim, D., Kim, K., Kang, M.: Electroless Ni–P-PTFE composite plating with rapid deposition and high PTFE concentration through a two-step process. Coatings **12** (2022). https://doi.org/10.3390/coatings12081199
24. Pratt, G.C.: Bearing materials: plain bearings. In: Encyclopedia of Materials: Science and Technology, pp. 488–496. Elsevier (2001)
25. Zainulabdeen, A.A., Hashim, F.A., Assi, S.H.: Mechanical properties of tin-based babbitt alloy using the direct extrusion technique. In: IOP Conference Series: Materials Science and Engineering. Institute of Physics Publishing (2019)
26. Guo, Q. qin, Guo, Y. chun, Yang, Z., Li, J. ping, Xu, G. tao: Study on microstructure and tribological properties of AlSn films deposited via magnetron sputtering ion plating. Appl. Surf. Sci. **483**, 123–132 (2019). https://doi.org/10.1016/j.apsusc.2019.03.233
27. Koring, R.: Changes in Plain Bearing Technology. SAE International (2013)
28. Adams, M.L.: Bearings. CRC Press, Boca Raton (2018)
29. Ouyang, J.-H., Li, Y.-F., Zhang, Y.-Z., Wang, Y.-M., Wang, Y.-J.: High-temperature solid lubricants and self-lubricating composites: a critical review. Lubricants **10**, 177 (2022). https://doi.org/10.3390/lubricants10080177
30. Renevier, N.M., Hamphire, J., Fox, V.C., Witts, J., Allen, T., Teer, D.G.: Advantages of using self-lubricating, hard, wear-resistant MoS-based coatings. 2 (2001)
31. Amaro, R.I., Martins, R.C., Seabra, J.O., Renevier, N.M., Teer, D.G.: Molybdenum disulphide/titanium low friction coating for gears application. Tribol. Int. **38**, 423–434 (2005). https://doi.org/10.1016/j.triboint.2004.09.003
32. Scharf, T.W., Kotula, P.G., Prasad, S.V.: Friction and wear mechanisms in MoS_2Sb_2O 2Au nanocomposite coatings. Acta Mater. **58**, 4100–4109 (2010). https://doi.org/10.1016/j.actamat.2010.03.040
33. Wahl, K.J., Sawyer, W.G.: Observing interfacial sliding processes in solid-solid contacts. MRS Bull. **33**, 1159–1167 (2008). https://doi.org/10.1557/mrs2008.246
34. Cao, H., De Hosson, J.T.M., Pei, Y.: Self-healing of a pre-notched WS2/a-C coating. Mater Res Lett. **7**, 103–109 (2019). https://doi.org/10.1080/21663831.2018.1561538
35. Cao, H., Bai, M., Inkson, B.J., Zhong, X., De Hosson, J.T.M., Pei, Y., Xiao, P.: Self-healing WS2 tribofilms: an in-situ appraisal of mechanisms. Scr. Mater. **204** (2021). https://doi.org/10.1016/j.scriptamat.2021.114124
36. Aouadi, S.M., Gu, J., Berman, D.: Self-healing ceramic coatings that operate in extreme environments: a review. J. Vacuum Sci. Technol. A: Vacuum, Surf. Films **38** (2020). https://doi.org/10.1116/6.0000350
37. Shirani, A., Gu, J., Wei, B., Lee, J., Aouadi, S.M., Berman, D.: Tribologically enhanced self-healing of niobium oxide surfaces. Surf. Coat. Technol. **364**, 273–278 (2019). https://doi.org/10.1016/j.surfcoat.2019.03.002
38. Erdemir, A., Ramirez, G., Eryilmaz, O.L., Narayanan, B., Liao, Y., Kamath, G., Sankaranarayanan, S.K.R.S.: Carbon-based tribofilms from lubricating oils. Nature **536**, 67–71 (2016). https://doi.org/10.1038/nature18948
39. Shirani, A., Li, Y., Eryilmaz, O.L., Berman, D.: Tribocatalytically-activated formation of protective friction and wear reducing carbon coatings from alkane environment. Sci. Rep. **11** (2021). https://doi.org/10.1038/s41598-021-00044-9
40. Stone, D., Liu, J., Singh, D.P., Muratore, C., Voevodin, A.A., Mishra, S., Rebholz, C., Ge, Q., Aouadi, S.M.: Layered atomic structures of double oxides for low shear strength at high temperatures. Scr. Mater. **62**, 735–738 (2010). https://doi.org/10.1016/j.scriptamat.2010.02.004
41. Franz, R., Mitterer, C.: Vanadium containing self-adaptive low-friction hard coatings for high-temperature applications: a review (2013)
42. Veprek, S., Argon, A.S.: Mechanical properties of superhard nanocomposites. Surf. Coat. Technol. **146–147**, 175–182 (2001). https://doi.org/10.1016/S0257-8972(01)01467-0

43. Baker, C.C., Chromik, R.R., Wahl, K.J., Hu, J.J., Voevodin, A.A.: Preparation of chameleon coatings for space and ambient environments. Thin Solid Films **515**, 6737–6743 (2007). https://doi.org/10.1016/j.tsf.2007.02.005
44. Fox-Rabinovich, G.S., Veldhuis, S.C., Dosbaeva, G.K., Yamamoto, K., Kovalev, A.I., Wainstein, D.L., Gershman, I.S., Shuster, L.S., Beake, B.D.: Nanocrystalline coating design for extreme applications based on the concept of complex adaptive behavior. J. Appl. Phys. **103** (2008). https://doi.org/10.1063/1.2904907
45. Shirani, A., Joy, T., Rogov, A., Lin, M., Yerokhin, A., Mogonye, J.E., Korenyi-Both, A., Aouadi, S.M., Voevodin, A.A., Berman, D.: PEO-Chameleon as a potential protective coating on cast aluminum alloys for high-temperature applications. Surf. Coat. Technol. **397** (2020). https://doi.org/10.1016/j.surfcoat.2020.126016
46. Fox-Rabinovich, G.: Adaptive hard coatings design based on the concept of self-organization during friction. In: Encyclopedia of Tribology, pp. 16–23. Springer US, Boston, MA (2013)
47. Voevodin, A.A., Zabinski, J.S.: Supertough wear-resistant coatings with 'chameleon' surface adaptation. Thin Solid Films **370**, 223–231 (2000). https://doi.org/10.1016/S0040-6090(00)00917-2
48. Voevodin, A.A., Muratore, C., Aouadi, S.M.: Hard coatings with high temperature adaptive lubrication and contact thermal management: review. Surf. Coat. Technol. **257**, 247–265 (2014). https://doi.org/10.1016/j.surfcoat.2014.04.046
49. Zhang, Y., Wang, Z., Zhou, S., Zhang, Y., Dong, Y., Wang, A., Ke, P.: Synergistic effect of V and Ag diffusion favored the temperature-adaptive tribological behavior of VAlN/Ag multi-layer coating. Tribol. Int. **192** (2024). https://doi.org/10.1016/j.triboint.2024.109285
50. Gachot, C., Rosenkranz, A., Hsu, S.M., Costa, H.L.: A critical assessment of surface texturing for friction and wear improvement. Wear **372–373**, 21–41 (2017). https://doi.org/10.1016/j.wear.2016.11.020
51. Vishnoi, M., Kumar, P., Murtaza, Q.: Surface texturing techniques to enhance tribological performance: a review. Surf. Interfaces **27** (2021). https://doi.org/10.1016/j.surfin.2021.101463
52. Rosenkranz, A., Costa, H.L., Baykara, M.Z., Martini, A.: Synergetic effects of surface texturing and solid lubricants to tailor friction and wear—a review. Tribol. Int. **155** (2021). https://doi.org/10.1016/j.triboint.2020.106792
53. He, D., He, C., Li, W., Shang, L., Wang, L., Zhang, G.: Tribological behaviors of in-situ textured DLC films under dry and lubricated conditions. Appl. Surf. Sci. **525** (2020). https://doi.org/10.1016/j.apsusc.2020.146581
54. Dorner-Reisel, A., Engel, A., Svoboda, S., Schürer, C., Weißmantel, S.: Laser structuring of hydrogenated DLC scaffolds: Raman spectroscopy and nanotribology. Diam. Relat. Mater. **108** (2020). https://doi.org/10.1016/j.diamond.2020.107787
55. Lu, C., Shi, P., Yang, J., Jia, J., Xie, E., Sun, Y.: Effects of surface texturing on the tribological behaviors of PEO/PTFE coating on aluminum alloy for heavy-load and long-performance applications. J. Mater. Res. Technol. **9**, 12149–12156 (2020). https://doi.org/10.1016/j.jmrt.2020.09.008
56. Chen, L., Liu, Z., Shen, Q.: Enhancing tribological performance by anodizing micro-textured surfaces with nano-MoS_2 coatings prepared on aluminum-silicon alloys. Tribol. Int. **122**, 84–95 (2018). https://doi.org/10.1016/j.triboint.2018.02.039
57. Pettersson, U., Jacobson, S.: Influence of surface texture on boundary lubricated sliding contacts. Tribol. Int. 857–864 (2003)
58. Rapoport, L., Moshkovich, A., Perfilyev, V., Lapsker, I., Halperin, G., Itovich, Y., Etsion, I.: Friction and wear of MoS_2 films on laser textured steel surfaces. Surf. Coat. Technol. **202**, 3332–3340 (2008). https://doi.org/10.1016/j.surfcoat.2007.12.009
59. Narayana, T., Saleem, S.S.: Enhancing fretting wear behavior of Ti64 alloy: The impact of surface textures and CrN–MoS_2–Ag composite coating. Tribol. Int. **193** (2024). https://doi.org/10.1016/j.triboint.2024.109346
60. Li, J., Zhang, X., Wang, J., Li, H., Huang, J., Xiong, D.: Frictional properties of silver overcoated on surface textured tantalum interlayer at elevated temperatures. Surf. Coat. Technol. **365**, 189–199 (2019). https://doi.org/10.1016/J.SURFCOAT.2018.10.067

References

61. Bondarev, A., Simonovic, K., Vitu, T., Kožmín, P., Syrovatka, Š., Polcar, T.: Textured coating or coated texture: femtosecond laser texturing of a-C:H/WC coatings for dry friction applications. Surf/ Coat Technol. **469** (2023). https://doi.org/10.1016/j.surfcoat.2023.129808

Chapter 3
DLC (Diamond-Like Carbon Coatings)

Abstract The chapter provides a concise characterisation of various forms of carbon coatings, along with fundamental information that is indispensable for further comprehension and elucidation of the phenomena concomitant with their tribological processes. The structural and chemical composition of carbon coatings are addressed. The elementary methods of their production and their effect on the resultant mechanical and tribological properties are also presented.

Carbon is the sixth most abundant element on Earth, with a prevalence of up to 94% in all known substances. It is widely regarded as one of the most significant building blocks of numerous chemicals, antibiotics, food products, and engineering and construction materials, which are indispensable to our healthy and modern lifestyles. This remarkable element, and in particular the multitude of varieties and forms it can take, offers a plethora of intriguing yet rare properties that expand its range of applications as science and technology advance. Examples include graphite, which is characterised by low hardness and excellent electrical conductivity and lubrication properties. In contrast, other carbon compounds can exhibit high hardness and toughness, such as transition metal carbides and natural and synthetic diamonds, which possess exceptional optical and semiconductor properties, in addition to mechanical properties. The wide range of crystalline and disordered carbon structures is mainly due to the existence of three different types of hybridisation of the electron orbitals that form the bonds between carbon atoms: sp^3, sp^2 and sp^1 (Fig. 3.1). It is due to this diversity it can manifest in a variety of forms, including solids such as graphite or diamond, as well as ordered structures, including fullerenes, nanotubes, nanofibres, nanopowders, nanocubes and thin films [1–4]. The latter can undoubtedly be considered a milestone in the development of civilisation, the advancement of which has often been determined by the availability of materials, in particular those possessing above-average and exceptional properties. The wide range of exceptional physical, mechanical, biomedical or tribological properties of coatings whose basic constituent is carbon undoubtedly makes this element a material that deserves the designation "exceptional".

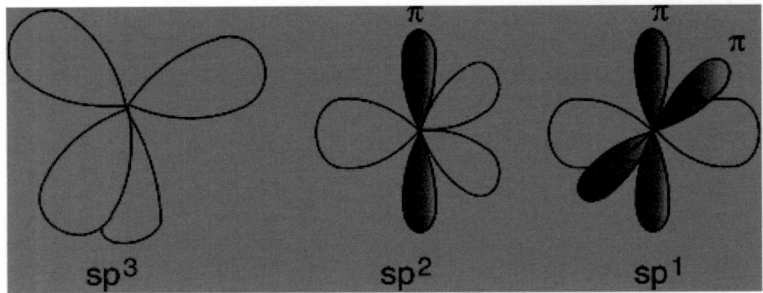

Fig. 3.1 Schematic representation of sp^3, sp^2 and sp^1 hybrid orbitals. Reprinted with permission from [2]

Carbon is utilised in the synthesis of two distinct categories of coatings: namely, diamond and transition metal carbide, nitride or boron carbide-based coatings. However, amorphous diamond-like carbon (DLC) coatings have gained the most popularity, mainly due to their unique combination of hardness, chemical resistance, low coefficient of friction and high wear resistance. These coatings were first synthesised in the 1970s by Aisenberg and Chabot [5], who used an ion beam to create coatings with diamond-like properties. The coatings obtained exhibited high hardness, modulus and resistivity, in addition to chemical inertness, a characteristic of diamond. This combination of properties led to the designation of these coatings diamond-like.

At present, these coatings can be produced using a variety of physical and chemical vapour deposition techniques, with different carbon sources employed as substrates. The properties and structural composition of the coatings obtained vary considerably depending on the deposition method used and the parameters of the synthesis process. Given the extensive literature on the synthesis and properties of carbon coatings produced by PVD and CVD methods [2, 6, 7], the author has taken the liberty of limiting this section to a brief characterisation of their different variants, as well as basic information of fundamental importance for the understanding and description of the phenomena accompanying their tribological processes. We shall therefore revert to the fundamental parameter that characterises amorphous carbon coatings, namely their structure. Materials with an amorphous structure are distinguished by the absence of long-range order, and consequently, they do not form a crystal lattice with a defined elemental cell type and parameters, as is the case with diamond or graphite. They exhibit short-range ordering, which manifests itself in a regular and predictable arrangement of atoms (as in a crystal lattice) over a small area, typically within one or two interatomic distances. However, deviations in lattice parameters are not uncommon when compared to the crystallographic lattice of diamond or graphite. The chemical structure of an amorphous carbon coating is illustrated alongside the elemental cells of diamond and graphite in Fig. 3.2.

So what determines the properties of DLC coatings? They are chiefly dictated by the content of sp^3 and sp^2 hybridized bonds in their chemical structure [2]. The sp^3

3 DLC (Diamond-Like Carbon Coatings)

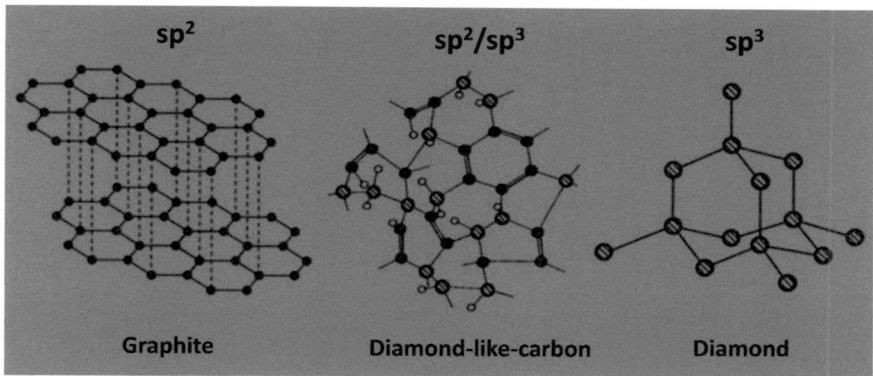

Fig. 3.2 Different hybridizations of carbon in diamond-like coatings. Reprinted with permission under Creative Common CC BY license from [8]

configuration is characteristic of a diamond. In this configuration, one s-orbital and three p-orbitals combine to form four identical sp^3 hybrid orbitals that point towards the corners of the tetrahedron, allowing them to form four strong σ bonds with neighbouring carbon atoms. This hybridisation gives diamond a three-dimensional structure with high density and hardness, making it the hardest known material [9]. In contrast, the sp^2 configuration is characteristic of graphite, where one s-orbital and two p-orbitals combine to form three identical planar sp^2 orbitals in a single plane. In this way they can form three equivalent bonds with neighbouring atoms in the form of hexagonal rings. An unoccupied p-orbital oriented perpendicular to the bond plane is capable of participating in the formation of π bonds with neighbouring atoms. As in the case with MoS$_2$ coatings, the individual atomic planes in graphite are connected by van der Waals bonds. Under shear forces, these bonds break and the material deforms plastically by sliding. However, in contrast to molybdenum disulphide, graphite exhibits increased sliding intensity in high humidity and in the presence of adsorbed oxygen [1, 10]. It can thus be deduced that, in the most general terms, the high hardness, wear resistance and modulus of elasticity of DLC coatings are related to the presence of sp^3 hybridized bonds in their chemical structure, while their low coefficient of friction is due to the presence of sp^2 hybridized bonds. However, this relationship is not as simple as it might appear at first glance. Firstly, it should be noted that diamond-like coatings exist not only in amorphous carbon (a-C) structures, but also in hydrogenated (a-C:H) varieties. The reason for the presence of this element, and not another, in the chemical composition of carbon coatings lies in the nature of a number of technologies for their production, which often use gaseous hydrocarbons as a source of carbon atoms. The systematics of the varieties of diamond-like coatings, contingent on the hydrogen content and the proportion of bonds with sp^3 and sp^2 hybridisation, is delineated by the triple phase diagram proposed by Jacob and Moller and shown in Fig. 3.3 [11].

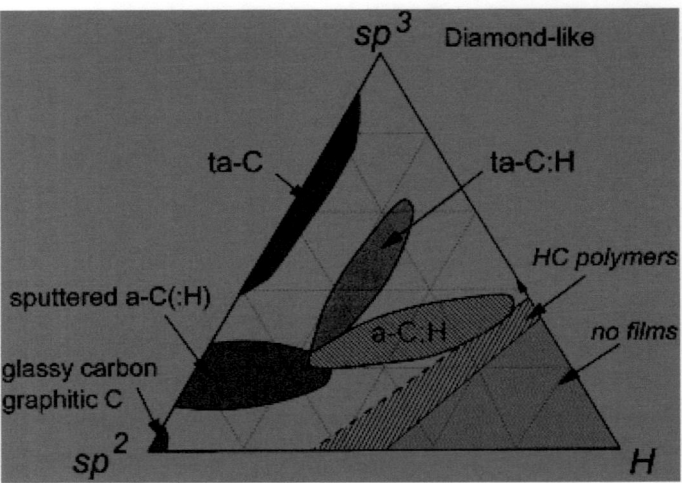

Fig. 3.3 Triple phase diagram of diamond-like carbon coatings. Reprinted with permission from [2]

From disordered a-C structures with predominantly sp^2 hybridised bonds to tough and wear resistant ta-C coatings with predominantly sp^3 hybridised bonds, the series along the left-hand side of the phase system includes coatings produced mainly by PVD *(Physical Vapour Deposition)* methods [12]. The a-C coatings with a low concentration of sp^3 bonded electrons are mainly obtained by classical magnetron sputtering *(DCMS—Direct Current Magnetron Sputtering)*, which is characterised by a relatively low degree of ionisation of the gaseous atmosphere [13, 14]. As magnetron sputtering technology has advanced over the years, new developments have emerged to take advantage of the highly ionised plasma. Innovative glow discharge excitation or polarisation systems for modified substrates have made it possible to produce carbon coatings with higher sp^3 (ta-C) hybrid bonding content. Revolutionary solutions in this field are the HiPIMS *(High Power Impulse Magnetron Sputtering)* [15–17] and GIMS *(Gas Injection Magnetron Sputtering)* [18, 19] methods. Among the other PVD technologies used to synthesise carbon coatings with a much higher sp^3 hybridised bonds content, the CVA *(Cathodic Vacuum Arc)* and FCVA *(Filtered Cathodic Vacuum Arc)* arc evaporation processes should be mentioned first [6, 20, 21]. The use of hydrogen gas atmospheres in the synthesis of a-C and ta-C coatings makes it possible to obtain coatings from within the triangle of the phase system, designated a-C:H and ta-C:H, respectively. Coatings of this type are mainly produced using methods from the broad group of PE CVD *(Plasma Enhanced Chemical Vapour Deposition)* technologies, using gaseous hydrocarbons as the carbon source [22]. Hydrogenated forms of carbon coatings, especially those with a high content of bonds with sp^3 hybridisation (ta-C:H) in their chemical structure, contain less hydrogen compared to a-C:H coatings and have a lower content of sp^3 bonds compared to their

3 DLC (Diamond-Like Carbon Coatings)

Table 3.1 Classification of DLC films for industrial applications [23]

	sp^3 (%)	Density	H (at%)	Hardness (GPa)
a-C	10–50	1.4–1.7	<5	9–25
ta-C	50–90	2.6–3.5	<5	25–90
a-C:H	10–50	1.4–2.0	5–50	9–25
ta-C:H	50–90	2.0–2.6	5–50	9–25

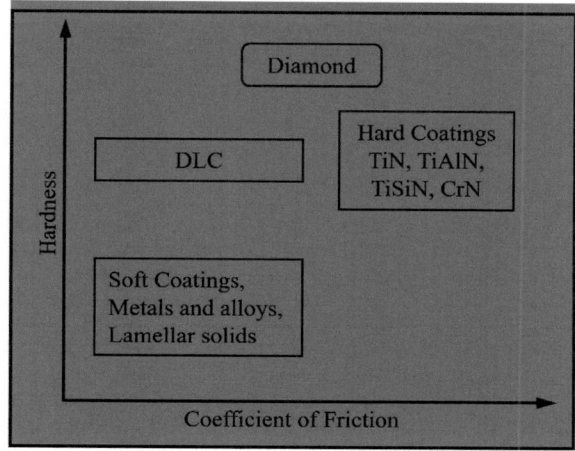

Fig. 3.4 Schematic representation of hardness and friction coefficient of carbon based coatings

non-hydrogenated type. A comparison of the different types of carbon coating in terms of composition and chemical structure and hardness is shown in Table 3.1.

The structure and chemical composition and resulting performance characteristics of carbon coatings are contingent on the technology and synthesis parameters employed, as well as the substrate material to be modified. Conventional PVD and PECVD processes frequently encounter limitations in their applicability due to high process temperatures and pressures, the inability to obtain homogeneous and uniform coatings on substrates with complex shapes, or production throughput limitations. Consequently, there is a need for systematic research to optimise the parameters of conventional and non-conventional deposition technologies in terms of their ability to produce DLC coatings at low process temperatures and atmospheric pressure, with production scalability to meet emerging market requirements [24].

As illustrated in Fig. 3.4, the dependence of hardness on the coefficient of friction varies significantly across different coating types, including hard carbide and nitride coatings, superhard nanocomposite coatings, and various types of carbon coatings.

The outcomes of studies, advancements and implementations in the domain of surface refinement of tools [25–29], machine and engine components [30–32] or medical implants [33–35] with DLC coatings provide compelling evidence for the considerable success of materials engineering in the past century. The extensive and clearly differentiated range of applications of carbon coatings renders them both scientifically fascinating and commercially crucial for numerous future industrial applications.

References

1. Hugh, O.: Pierson: Handbook of Carbon, Graphite, Diamond and Fullerens: Properties. Noyes Publications, Processing and Applications (1994)
2. Robertson, J.: Diamond-like amorphous carbon. Mater. Sci. Eng. R. Rep. **37**, 129–281 (2002). https://doi.org/10.1016/S0927-796X(02)00005-0
3. Nanodiamonds. Elsevier (2017)
4. Ho, D. (ed.): Nanodiamonds. Springer, US, Boston, MA (2010)
5. Aisenberg, S., Chabot, R.: Ion-beam deposition of thin films of diamondlike carbon. J. Appl. Phys. **42**, 2953–2958 (1971). https://doi.org/10.1063/1.1660654
6. Vetter, J.: 60 years of DLC coatings: historical highlights and technical review of cathodic arc processes to synthesize various DLC types, and their evolution for industrial applications. Surf. Coat. Technol. **257**, 213–240 (2014). https://doi.org/10.1016/j.surfcoat.2014.08.017
7. Milton, O.: Materials Science of Thin Films. Elsevier (2002)
8. Grigoriev, S.N., Volosova, M.A., Fedorov, S.V., Mosyanov, M.: Influence of DLC coatings deposited by PECVD technology on the wear resistance of carbide end mills and surface roughness of AlCuMg$_2$ and 41Cr$_4$ Workpieces. Coatings **10**, 1038 (2020). https://doi.org/10.3390/coatings10111038
9. Field, J.E. (ed.): The Properties of Natural and Synthetic Diamond. Academic Press, London (1993)
10. Scharf, T.W., Prasad, S.V.: Solid lubricants: a review. J. Mater. Sci. **48**, 511–531 (2013). https://doi.org/10.1007/s10853-012-7038-2
11. Jacob, W., Möller, W.: On the structure of thin hydrocarbon films. Appl. Phys. Lett. **63**, 1771–1773 (1993). https://doi.org/10.1063/1.110683
12. Mattox, D.M.: Handbook of Physical Vapor Deposition (PVD) Processing. Elsevier (2010)
13. Christou, C., Barber, Z.H.: Ionization of sputtered material in a planar magnetron discharge. J. Vac. Sci. Technol. A: Vac. Surf. Films **18**, 2897–2907 (2000). https://doi.org/10.1116/1.1312370
14. Huang, M., Zhang, X., Ke, P., Wang, A.: Graphite-like carbon films by high power impulse magnetron sputtering. Appl. Surf. Sci. **283**, 321–326 (2013). https://doi.org/10.1016/j.apsusc.2013.06.109
15. Claver, A., Jiménez-Piqué, E., Palacio, J.F., Almandoz, E., De Ara, J.F., Fernández, I., Santiago, J.A., Barba, E., García, J.A.: Comparative study of tribomechanical properties of HiPIMS with positive pulses DLC coatings on different tools steels. Coatings **11**, 1–21 (2021). https://doi.org/10.3390/coatings11010028
16. García, J.A., Rivero, P.J., Barba, E., Fernández, I., Santiago, J.A., Palacio, J.F., Fuente, G.G., Rodríguez, R.J.: A comparative study in the tribological behavior of DLC coatings deposited by HiPIMS technology with positive pulses. Metals (Basel) **10** (2020). https://doi.org/10.3390/met10020174
17. Santiago, J.A., Fernández-Martínez, I., Kozák, T., Capek, J., Wennberg, A., Molina-Aldareguia, J.M., Bellido-González, V., González-Arrabal, R., Monclús, M.A.: The influence of positive pulses on HiPIMS deposition of hard DLC coatings. Surf. Coat. Technol. **358**, 43–49 (2019). https://doi.org/10.1016/j.surfcoat.2018.11.001
18. Chodun, R., Skowronski, L., Trzcinski, M., Nowakowska-Langier, K., Kulikowski, K., Naparty, M., Radziszewski, M., Zdunek, K.: The amorphous carbon thin films synthesized by gas injection magnetron sputtering technique in various gas atmospheres. Coatings **13** (2023). https://doi.org/10.3390/coatings13050827

References

19. Wicher, B., Chodun, R., Kwiatkowski, R., Trzcinski, M., Nowakowska-Langier, K., Lachowski, A., Minikayev, R., Rudnicki, J., Naparty, M.K., Zdunek, K.: Plasmochemical investigations of DLC/WCx nanocomposite coatings synthesized by gas injection magnetron sputtering technique. Diam. Relat. Mater. **96**, 1–10 (2019). https://doi.org/10.1016/J.DIAMOND.2019.04.025
20. Kang, M.C., Tak, H.S., Jeong, Y.K., Lee, H.W., Kim, J.S.: Properties and tool performance of ta-C films deposited by double-bend filtered cathodic vacuum arc for micro drilling applications. Diam. Relat. Mater. **19**, 866–869 (2010). https://doi.org/10.1016/j.diamond.2010.02.002
21. Poplavsky, A.I., Ya. Kolpakov, A., Galkina, M.E., Kovaleva, M.G., Japrintsev, M.N., Mishunin, M.V., Gerus, J.V.: Properties of carbon coatings obtained by pulsed high power methods of vacuum-arc and magnetron sputtering. Mater. Today Proc. **5**, 25933–25938 (2018). https://doi.org/10.1016/j.matpr.2018.08.006
22. Pierson, H.O.: Handbook of Chemical Vapor Deposition. Elsevier (1992)
23. Ohtake, N., Hiratsuka, M., Kanda, K., Akasaka, H., Tsujioka, M., Hirakuri, K., Hirata, A., Ohana, T., Inaba, H., Kano, M., Saitoh, H.: Properties and classification of diamond-like carbon films. Materials **14**, 315 (2021). https://doi.org/10.3390/ma14020315
24. Zia, A.W., Birkett, M.: Deposition of diamond-like carbon coatings: conventional to non-conventional approaches for emerging markets. Ceram. Int. **47**, 28075–28085 (2021). https://doi.org/10.1016/j.ceramint.2021.07.005
25. Fukui, H., Okida, J., Omori, N., Moriguchi, H., Tsuda, K.: Cutting performance of DLC coated tools in dry machining aluminum alloys. Surf. Coat. Technol. **187**, 70–76 (2004). https://doi.org/10.1016/j.surfcoat.2004.01.014
26. Martins, P.S., Almeida Magalhães Júnior, P.A., Gonçalves Carneiro, J.R., Talibouya Ba, E.C., Vieira, V.F.: Study of diamond-like carbon coating application on carbide substrate for cutting tools used in the drilling process of an Al–Si alloy at high cutting speeds. Wear **498–499**, 204326 (2022). https://doi.org/10.1016/j.wear.2022.204326
27. Jatti, V.S., Sefene, E.M., Jatti, A.V., Mishra, A., Dhabale, R.D.: Synthesis and characterization of diamond-like carbon coatings for drill bits using plasma-enhanced chemical vapor deposition. Int. J. Adv. Manuf. Technol. **127**, 4081–4096 (2023). https://doi.org/10.1007/s00170-023-11794-3
28. Kowalczyk, J., Madej, M., Milewski, K., Nowakowski, Ł., Ozimina, D.: The Influence of Cutting Fluid and Diamond-Like Carbon Coating on Cutting Tool Wear (2019)
29. Wang, C., Shan, D., Guo, B.: Chapter 21—DLC-coated tools for micro-forming. In: Qin, Y. (ed.) Micromanufacturing Engineering and Technology, 2nd edn., pp. 487–512. William Andrew Publishing, Boston (2015)
30. Mutyala, K.C., Singh, H., Evans, R.D., Doll, G.L.: Effect of diamond-like carbon coatings on ball bearing performance in normal, oil-starved, and debris-damaged conditions. Tribol. Trans. **59**, 1039–1047 (2016). https://doi.org/10.1080/10402004.2015.1131349
31. Hauert, R.: An overview on the tribological behavior of diamond-like carbon in technical and medical applications. Tribol. Int. **37**, 991–1003 (2004). https://doi.org/10.1016/j.triboint.2004.07.017
32. Humphrey, E., Elisaus, V., Rahmani, R., Mohammadpour, M., Theodossiades, S., Morris, N.: Diamond like-carbon coatings for electric vehicle transmission efficiency. Tribol. Int. **189**, 108916 (2023). https://doi.org/10.1016/j.triboint.2023.108916
33. Hauert, R., Falub, C.V., Thorwarth, G., Thorwarth, K., Affolter, C., Stiefel, M., Podleska, L.E., Taeger, G.: Retrospective lifetime estimation of failed and explanted diamond-like carbon coated hip joint balls. Acta Biomater. **8**, 3170–3176 (2012). https://doi.org/10.1016/j.actbio.2012.04.016
34. Hauert, R., Thorwarth, K., Thorwarth, G.: An overview on diamond-like carbon coatings in medical applications. Surf. Coat. Technol. **233**, 119–130 (2013). https://doi.org/10.1016/j.surfcoat.2013.04.015
35. Thin Film Coatings for Biomaterials and Biomedical Applications. Elsevier (2016)

Chapter 4
Tribological Properties of DLC Coatings

Abstract The chapter provides a concise overview of the origins of tribological behaviour of diamond-like coatings. It presents an analysis of the external and internal factors that affect the friction coefficient and wear resistance. The chapter pays particular attention to such elements as geometric surface structure, passivation of dangling bonds, the role of tribo-induced graphitisation and formation of transfer layer.

The combination of high chemical inertness, excellent wear resistance and ultra-low coefficients of friction achieved by carbon coatings makes them a unique material from a tribological point of view, capable of meeting increasingly stringent performance requirements. These properties have found their way into a number of demanding applications and have become the subject of excellent scientific studies, for example by Hauert [1, 2], Erdemir [3, 4], Donnet and Erdemir [5], or Homberg and Matthews [6, 8, 9] and Matthews et al. [7]. The broad array of forms and types of DLC's, in conjunction with the extensive range of test conditions and other environmental factors, gives rise to significant variations in the quantitative description of their tribological properties, as reported in the extant literature. The hardness, coefficient of friction and wear resistance of DLC coatings are contingent on the synthesis technology, substrates utilised and deposition parameters, and are thus deemed relative quantities. The range and magnitude of these properties make them the material with the widest known spectrum of tribological properties. In general, a wear resistance of $<10^{-11}$ mm^3 N^{-1} m^{-1} and a coefficient of friction in the range of 0.001–0.7, obtained under dry friction conditions, are most characteristic of carbon films. The factors influencing such a wide range of tribological properties can be divided into two groups:

internal factors: depending on the mutual concentration of sp^2/sp^3 hybridised carbon bonds in the amorphous matrix, the content of hydrogen or other dopant atoms (sometimes such elements are deliberately added during the coating synthesis process to modify their functional properties), as well as properties resulting directly from the synthesis technology, the substrates used or the parameters of the deposition process. Figure 4.1 illustrates the interdependence of the service and performance properties of DLC coatings as a function of the parameters employed and the synthesis technology.

© The Author(s), under exclusive license to Springer Nature Switzerland AG 2025
D. Batory, *Tribology of Low Friction Carbon Based Coatings*, Engineering Materials,
https://doi.org/10.1007/978-3-031-95979-0_4

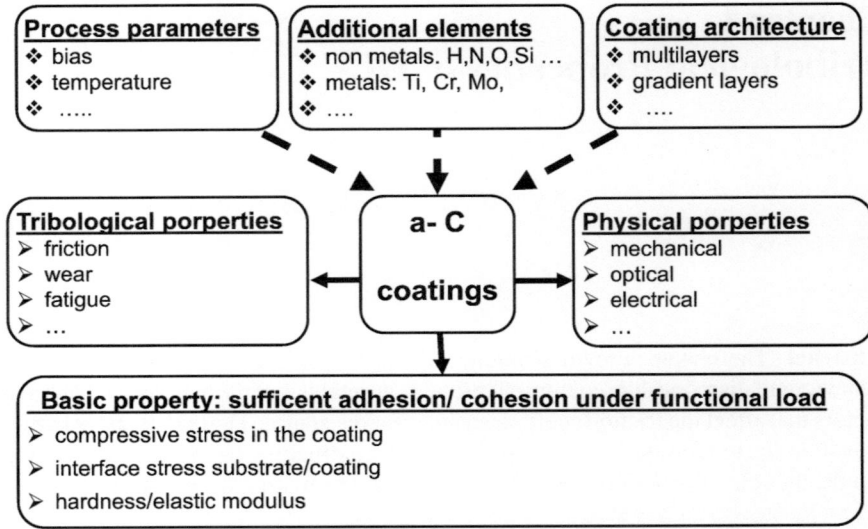

Fig. 4.1 Methods for tailoring functional properties of DLC coatings [10, 11]

external factors: those resulting from physical, chemical or mechanical interactions of the coating material with the counterbody and the environment. The impact of individual external factors on overall tribological properties is largely contingent upon the specific test conditions and parameters employed. The following factors warrant particular consideration: the geometrical structure of the substrate surface and the counterbody material; the counterbody material itself; the applied load; the type of cooperation; the sliding speed; the sliding distance; humidity; and the composition of the working atmosphere.

4.1 The Influence of the Surface Geometrical Structure of DLC Coatings on Their Tribological Properties

The combination of materials with a high values of surface roughness almost always leads to an increase in frictional losses. The mutual mechanical blocking of the tips of the irregularities on the mating surfaces results in an increase in the coefficient of friction and a rapid progression of wear, particularly during the initial phase, known as the running-in period. Due to their amorphous structure, carbon coatings are able to reproduce the parameters of the geometric structure of the substrate surface with great accuracy, and in some cases even reduce them. It can therefore be postulated that the frictional resistance component resulting from the mechanical interactions of the roughness peaks of the coating will, in almost all cases, have the same or a lower value compared to the unmodified substrate. Figure 4.2 shows a comparison of the

4.1 The Influence of the Surface Geometrical Structure of DLC Coatings …

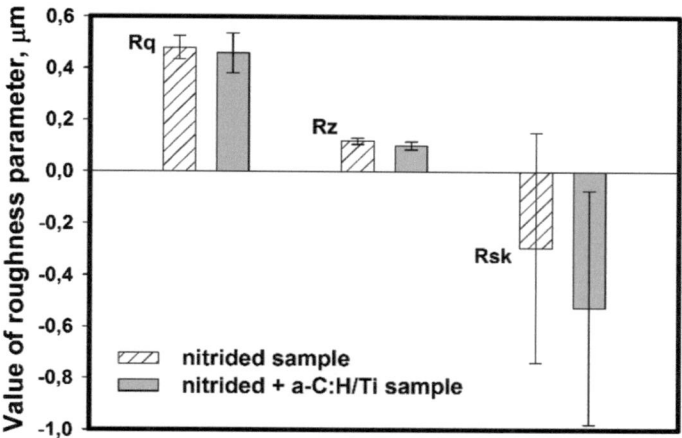

Fig. 4.2 Effect of carbon coating on surface roughness parameters. Reprinted with permission under Creative Commons license form [12]

surface roughness parameters of nitrided H9S2 steel before and after modification with a 1 μm thick DLC coating.

The Rq parameter is defined as the surface slope coefficient, also known as kurtosis. The Rz parameter is defined as the average value of the height of the roughness profile, while the Rsk parameter is defined as the surface asymmetry coefficient, also known as skewness. With regard to the Rz parameter, a slight decrease in its value is observed, indicating a reduction in the average distance between the highest elevation and the lowest depression of the profile. Conversely, the use of a surface with a higher kurtosis and greater negative skewness always results in a lower frictional force and a shorter sliding distance to the steady state, although there is no difference in the average roughness. The topography of a plateau-type contact surface with high kurtosis and more negative skewness, and the effect of kurtosis and skewness on the value of the coefficient of friction, indicate that these parameters can be used as design parameters to modify surface topography and texture with the aim of reducing the frictional force [13]. It should be noted that some of the technologies used in the synthesis of DLC coatings, in particular arc evaporation techniques, are associated with the release of droplets of liquid material from the surface of the carbon substrate undergoing evaporation. These droplets are characterised by diameters ranging from hundreds of parts per thousand to tens of microns. During the synthesis process, the droplets are transferred to the modified substrate and incorporated into the growing coating. This results in the formation of irregular structures and an increase in surface roughness parameters, which can lead to an increase in the coefficient of friction and wear in the initial phase of friction node cooperation [14, 15]. A number of treatments have been used to mitigate this adverse phenomenon, including electromagnetic separation of microdroplets from arc discharge plasma [16, 17]. Figure 4.3 shows a schematic of the FCVA system, while Fig. 4.4 shows the surface structure of the DLC coating produced by the CVA process without additional microdroplet separation system.

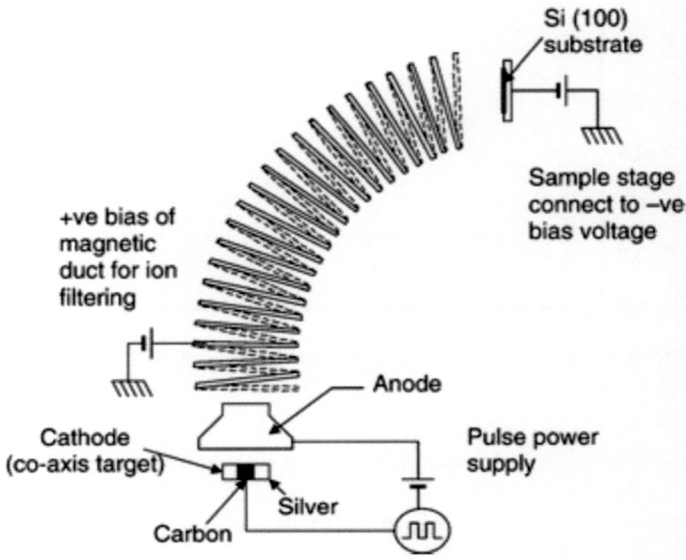

Fig. 4.3 Schematic of the FCVA system. Reprinted with permission form [16]

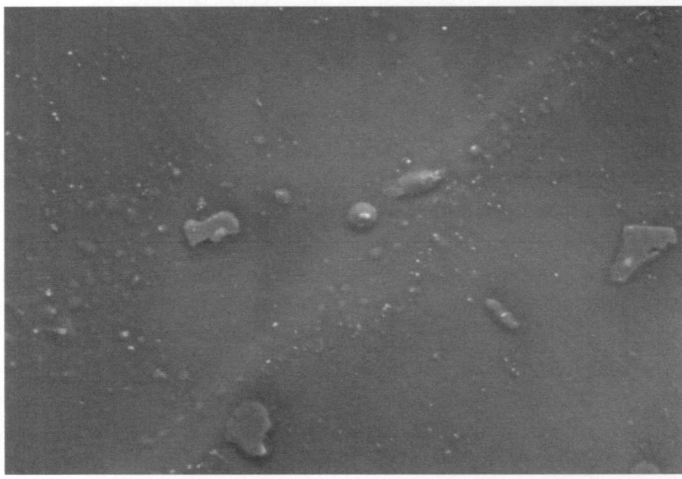

Fig. 4.4 SEM image of surface structure of diamond-like carbon layer deposited using vacuum arc technique. Courtesy of Dr. M. Makówka, Institute of Materials Science and Engineering TUL

4.2 Saturation of Dangling Bonds of Carbon Atoms at the Sliding Interface with Passivating Species (Passivation Mechanism) in the Tribology of Diamond-Like Carbon Coatings

One of the fundamental properties of diamond-like carbon coatings is their chemical inertness. This means that the surface of the coating is chemically stable and will not react chemically with other solids, liquids or gases under static conditions. However, this does not mean that the surface is free from adsorbed compounds or impurities. The synthesis processes of carbon coatings for both PVD and CVD technologies are carried out in specially selected and tightly controlled atmospheres. However, after the synthesis process, the surface, or more precisely the active centres, become occupied by adsorbed gas molecules (such as water vapour, oxygen, hydrocarbons, etc.) as a result of contact with the components of the atmosphere. As a result of the dynamic interactions that occur at the frictional interface, the adsorbed molecules are removed, allowing the free surface of the coating to interact with the surrounding chemicals and the countetnody material. Although the literature on this subject makes a distinction between interactions of an adhesive and tribochemical nature, the author believes that the two issues are interrelated and that a detailed delineation of them, together with the associated scientific basis is beyond the scope of this book. From a frictional point of view, it is essential to distinguish between the two issues due to the nature of the interaction involved. Specifically, this refers to the mutual interaction between the surfaces of the friction pairing elements, as well as their chemical interaction with the surrounding environment. As previously discussed, carbon coatings are characterised by a lack of crystallographic structure. In contrast, their highly disordered chemical structure consists mainly of a mixture of sp^3 and sp^2 hybridized bonds, i.e. covalent bonds. The spatial nature of the sp^3 bonds results in the formation of unsaturated bonds on the surface of the carbon film, which are capable of forming strong covalent bonds with other atoms. The energy of these bonds is the highest of any found in carbon materials, and therefore they are the cause of significant friction losses due to surface adhesion interactions at the friction interface. In accordance with the theoretical model of dry friction as proposed by Bowden and Tabor [18], it can be observed that, in addition to the mechanical interactions, the adhesion forces present in the areas of actual contact between the surfaces of the friction pairing elements constitute the second dominant component of the friction force. The initial and primary factor that significantly influences the nature of the coating's interaction with both the mating material and the surrounding environment is the hydrogen present in its chemical composition. Hydrogen has been observed to occur in two different forms: firstly, in carbon-bonded form, and secondly occupying the voids within the coating structure (see Fig. 4.5) [19].

The utilisation of hydrocarbons, or the deliberate introduction of hydrogen into the working chamber during the synthesis of diamond-like coatings, is a common and justifiable practice, primarily due to the low friction coefficient values and high wear resistance that are obtained [20]. The underlying principle of this phenomenon

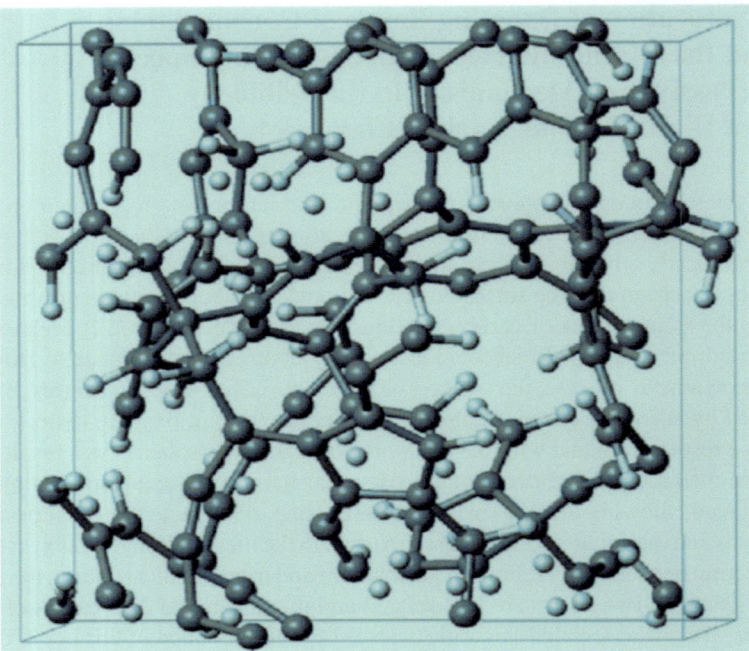

Fig. 4.5 Atomic scale simulation of a highly hydrogenated DLC film showing carbon–hydrogen random network structure with some free hydrogen occupying interstitial positions. Reprinted with permission under Creative Commons CC BY from [15]

is the high chemical affinity between carbon and hydrogen, whereby free σ-bonds on the surface of the coating are saturated by monovalent hydrogen atoms. This process effectively hinders their subsequent adhesive interaction with molecules such as oxygen, water vapour, and other gases present in the working atmosphere, which often exert a detrimental influence on the friction coefficient value [3, 21–23]. A comprehensive analysis of test results pertaining to hydrogenated and non-hydrogenated carbon coatings under reduced pressure, in controlled atmospheres or controlled humidity, has clearly demonstrated discrepancies in their tribological properties. The absence of adsorbed gas molecules or other functional groups on the surface of DLC coatings, which can interact with free σ-bonds or act as a solid lubricant, indicates that the composition and chemical structure of the coating are the primary factors determining its tribological properties (see Fig. 4.6 for a schematic representation of tribochemical interactions at the contact interface of carbon coatings).

It has been demonstrated that a-C:H coatings exhibit markedly reduced friction coefficients and enhanced wear resistance when subjected to conditions of reduced pressure (high vacuum) and in inert atmospheres (Ar, N_2, He), in comparison to their non-hydrogenated varieties. In the presence of a high vacuum, the compounds that typically occupy free σ-bonds on the surface of a-C coatings are subjected to

4.2 Saturation of Dangling Bonds of Carbon Atoms at the Sliding Interface ...

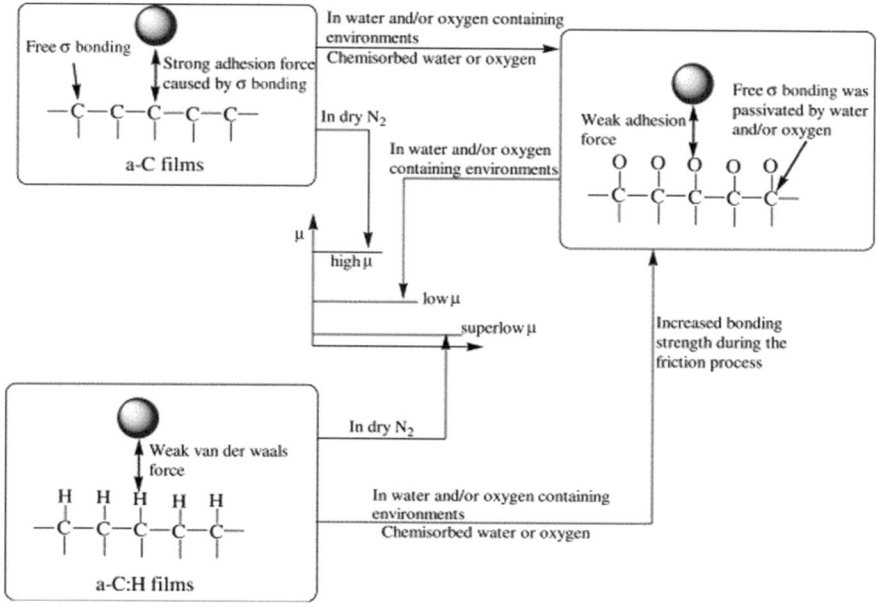

Fig. 4.6 Schematic illustration of the tribochemical effects on the friction behaviors of DLC films. Reprinted with permission from [24]

mechanical or thermal removal processes. This process enables the coating surface, which is inherently reactive, to form adhesive bonds with the counterbody material. Consequently, the coefficient of friction increases significantly, reaching values of the order of $\mu > 0.5$, and there is progressive wear of both the coating and the counterbody material. Under high vacuum conditions, hydrogen desorbs gradually from a-C:H coatings. This process has been shown to saturate the C–H surface bonds, which are continually destroyed throughout the tribological process, thereby effectively preventing the formation of adhesive bonds at the friction interface. In contrast, when both coating types are operated under controlled (high) relative humidity conditions, the opposite is observed. The free surface bonding of carbon atoms in the non-hydrogenated a-C coating is saturated by hydrogen and hydroxyl (–OH) groups derived from the dissociation of water vapour molecules present in the working atmosphere. Consequently, their capacity to form adhesive bonds with the counterbody material is significantly constrained, as evidenced by a notable reduction in the friction coefficient value. In contrast, hydrogenated a-C:H coatings are renowned for their sensitivity to changes in the friction coefficient, which is observed to increase during testing at relative humidities above 30%. The oxidation process that occur during testing result in the replacements of surface C–H bonds with oxygen-containing –OH groups, which exhibit a higher binding energy in the C–OH/C–OH contact zone (approximately 0.21 eV) when compared to Van der Waals C–H/C–H bonds (0.08 eV). This results in an increase in the friction coefficient and wear of the coating [25–27]. However, it appears that a specific relative humidity threshold exists, beyond which the amount of water adsorbed on the contact surface leads to an intensification of hydrogen–oxygen interactions, thereby attaining its maximum efficiency. With an increase in the relative humidity of the working atmosphere, the adsorbed water functions as a lubricant, thereby diminishing the hydrogen–oxygen interaction and contributing to a reduction in the friction coefficient [28]. The employment of controlled working atmospheres in the testing of a-C:H coatings (Ar, N_2) primarily indicates their protective nature. The inert gas functions as a barrier to residual amounts of moisture molecules, which, as a consequence of tribochemical reactions, could result in a reconfiguration of the C–H to C–OH surface bonds and, concomitantly, adversely affect the friction coefficient value. In the context of non-hydrogenated coatings, the employment of controlled humidity atmospheres comprising inert gases is particularly advantageous [24, 29]. Figure 4.7 provides a schematic representation of the interaction of unsaturated carbon bonds on the surface of a-C coating with a controlled working atmosphere, whereas Fig. 4.8 shows a range of friction mechanisms exhibited by diverse DLC films at the molecular level.

The outcomes of the friction coefficient assessments of DLC coatings in a controlled working atmosphere and varying values of the relative velocity of the friction contact demonstrate that the test parameters exert a significant influence on the resulting value. In contrast to the results obtained under normal conditions, the value of the friction coefficient of a-C:H coatings does not decrease in proportion to an increase in the sliding speed under reduced pressure. This phenomenon can be attributed to the rise in temperature of the friction node, in addition to the reduction in interaction time between the friction contact surface and the gaseous components of the working atmosphere. These phenomena result in a decrease in the number of gas molecules adsorbed on the surface, thereby favouring low friction coefficient values [31, 32]. In ultra-high vacuum conditions, the probability of tribochemical interactions between the active components of the working atmosphere and the friction surface is negligible. Consequently, it can be hypothesised that the friction coefficient value is primarily dependent on the dynamics of repassivation processes of free carbon bonds occupied by permanently removed hydrogen atoms, as well as other phenomena occurring directly in the chemical structure of the coating [33]. A more thorough examination of these phenomena is provided in the subsequent chapter. Recently, a novel concept has been proposed by Nakayama, which is based on tribocharging, tribomicroplasma generation and triboemission phenomena [34, 35].

4.3 The Role of Stress and Shear Induced (Tribo-Induced) Graphitization ... 79

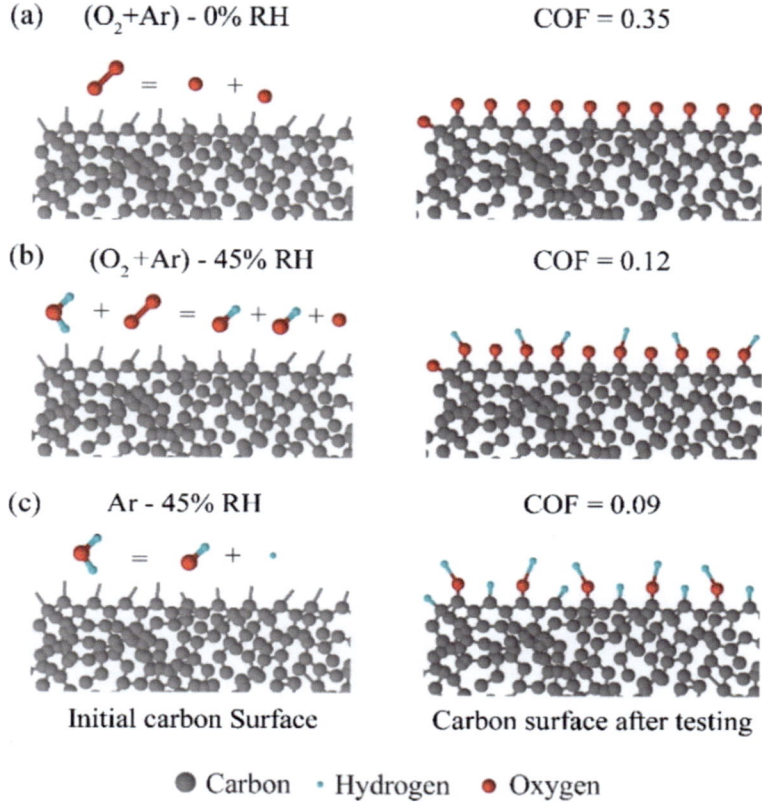

Fig. 4.7 Schematic representation of the surface termination mechanisms [27]

4.3 The Role of Stress and Shear Induced (Tribo-Induced) Graphitization in Tribology of Diamond-Like Carbon Coatings

The markedly low friction coefficient values observed in carbon coatings, particularly in their hydrogenated variants, have been demonstrated to correlate with the surface passivation of free carbon bonds by hydrogen atoms, as evidenced under controlled laboratory conditions. This assumption, although seemingly justified and supported by abundant evidence, does not, however, explain the fact of their excellent wear resistance, particularly under dry friction conditions. It can thus be concluded that the energy of the C–H bonds is an insufficient barrier to progressive wear processes, even in the case of a friction pair composed of surfaces modified with a carbon coating. The question thus arises as to how their ultra-high wear resistance can be explained. Given the metastable structure of amorphous carbon coatings, it is proposed that the type of carbon bond hybridization can be transformed into a more stable form, provided that

Type of DLC	a-C:H	a-C:H or a-C	H-free DLC (both ta-C and a-C)
Friction range	< 0.02	0.1 – 0.2	> 0.5
Nature of the interaction	Van der Waals	Hydrogen	σ or π
Energy (eV/bond)	0.08	0.2	0.4 – 0.8
Schematic of the interaction (environment)	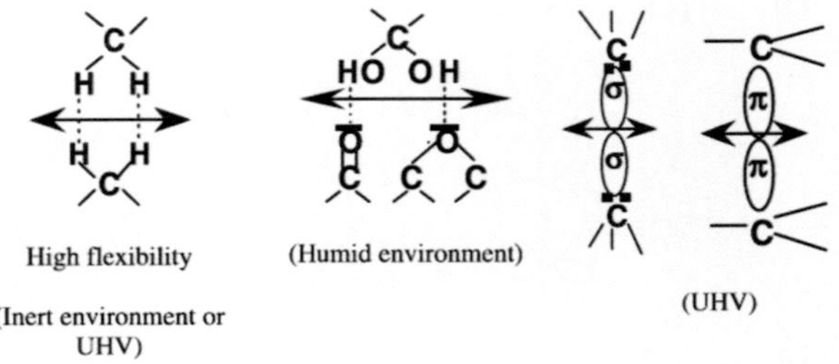 High flexibility (Inert environment or UHV)	(Humid environment)	(UHV)

Fig. 4.8 Friction mechanisms exhibited by diverse DLC films at the molecular level. Reprinted with permission from [30]

the energy barrier associated with a given phase transformation is overcome. This transformation is the local transition of carbon bonds with sp^3 hybridization to bonds with sp^2 hybridization. The tribochemical reactions occurring at the friction interface, including the breaking and formation of chemical bonds, the release of hydrogen, and the migration and regrouping of atoms at interfacial boundaries, result in the formation of a thin transfer layer with low shear resistance. This layer acts as a solid lubricant, protecting the surface of the component that is mating with it, reducing the wear rate of the layer itself, and leading to a lower friction coefficient [12, 36–38]. The phase transformation process can be induced by reaching the flash temperature at the contact points of the friction pair surface roughness asperities. The average increase in flash temperature caused by sliding friction occurring between the friction pair surface roughness vertices can be determined by the following equation [37]:

$$\Delta T = \frac{\mu P v}{4a(k_{DLC} + k_p)} \quad (4.1)$$

where:

ΔT temperature increase in the contact zone of the surface asperities
μ coefficient of friction
P applied load
v relative sliding speed
a radius of actual contact zone
k thermal conductivity coefficient of the DLC coating (K_{DLC}) and the counterbody (K_P).

4.3 The Role of Stress and Shear Induced (Tribo-Induced) Graphitization ...

The data demonstrate a direct proportionality between the local increase in contact temperature and the coefficient of friction, the load on the friction node and its relative speed, and an inverse proportionality between the local increase in contact temperature and the actual contact area. Therefore, even for relatively low values of the sliding speed and load on the friction node, a reduction in the actual contact area can result in the flash temperature exceeding the critical value of the phase transition temperature. Furthermore, an increase in the value of load and relative velocity of the friction pair at a fixed value of the friction coefficient will additionally lead to an intensification of the progressive coating graphitisation processes. The results of the research into the kinetics of transfer layer formation for friction nodes operating at high loads and high relative sliding speed in an ambient atmosphere indicate that both parameters favour the intensification of the graphitisation process. It has been demonstrated that an increase in load and sliding speed results in a decrease in the friction and wear coefficient values of the DLC coatings under investigation [37]. This finding indicates that as the Hertz stress in the contact zone increases, the coefficient of friction of DLC coatings decreases. This conclusion is supported by other literature reports on the study of the dependence of the friction coefficient of DLC coatings on the normal force in humid air and dry nitrogen atmospheres. The findings of this study align closely with the conclusions drawn from the application of the law of least squares, which indicates a high degree of correlation between the friction coefficient and the inverse of the friction node load value. This correlation is further supported by the findings of the Hertzian contact model, introduced in Chap. 2, Eq. 2.8 [39, 40]:

$$\mu = S_0 \pi \left(\frac{3R}{4E^*}\right)^{\frac{2}{3}} W^{-\frac{1}{3}} + \alpha \tag{4.2}$$

where:

S_0 interfacial shear strength between the transfer film and underlying wear track
R radius of ball
E^* is the combined plane stress modulus for the two surfaces
W applied load
α pressure dependence of shear strength, which is the lowest attainable friction coefficient for the tested couples.

In light of the fact that this theory contradicts Amonton's first law of friction—the independence of the coefficient of friction from the load on the friction node—this behaviour is referred to as non-Amontonian. Furthermore, this has been confirmed for other low-friction materials [39, 41].

As a consequence of progressive graphitisation processes, a thin (approximately 3 nm) transfer layer is formed in the surface areas of the carbon coating. This transfer layer has a locally ordered structure that resembles graphite planes aligned parallel to the surface [38, 42]. Figure 4.9c shows the transmission electron microscope image of a cross-section of the wear track in the surface of the carbon coating. The

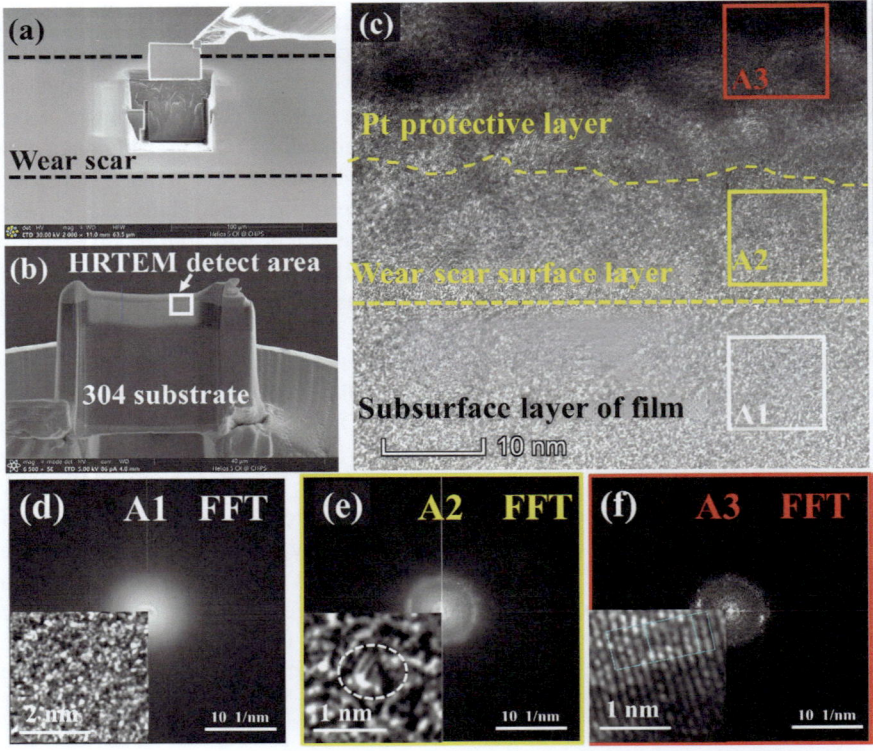

Fig. 4.9 Cross-sectional HRTEM of the wear track of DLC layer. Reprinted with permission under Creative Commons CC BY form [38]

Fourier transform of the electron diffraction images demonstrates the formation of nanocrystalline-sized ordered structures in the zone of direct friction pair interaction, thereby confirming the formation of a transition layer with sp^2 hybridisation (see Fig. 4.9e).

From the perspective of the scale of the phase transformations occurring at the frictional contact, their experimental analysis poses significant challenges, primarily due to the nanometre-scale dimensions of the observed volumes and the technical viability of implementing experimental techniques of solid-state physics for their analysis, given the ongoing nature of tribological processes. A significant and widely debated topic in this context is the fact that the majority of the associated theses are based on "post mortem" analysis of wear track on the surface of the coating and wear scar on the surface of the counterbody. The advent of computer technology has opened up a plethora of avenues for numerical modelling of friction-induced metastable transformations of amorphous carbon coatings. MD computer simulation techniques offer the potential to analyse frictional phenomena at any point of contact and at any time, thereby providing a more comprehensive insight into the dynamics of tribological processes. Furthermore, these techniques facilitate a deeper understanding of these phenomena. The outcomes of molecular dynamics computer

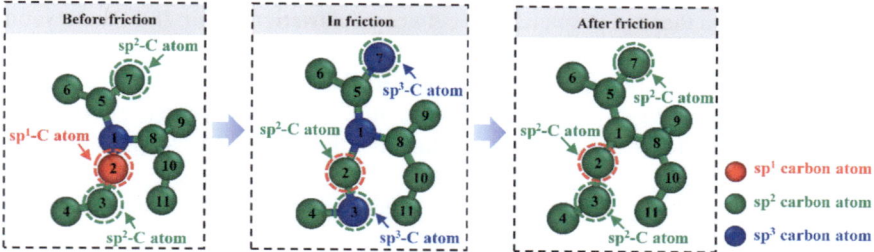

Fig. 4.10 The typical transformation way of C atom hybridization modes in the amorphous carbon film during the friction process. Reprinted with permission from [44]

simulations of the transformation processes of highly disordered structures with sp^3 hybridization, forming areas of increasing order and graphite- or graphene-like structures in the contact zone, indicate that the dominant mechanism and driving force behind these processes is stress relaxation under the action of shear forces [43]. Subsequent research in this field corroborates this hypothesis, yet introduces an additional element to the transformation. Specifically, as the authors indicate in a particular volume range, this transformation is reversible. The hybridization of bonds between carbon atoms undergoes a metastable transformation as a result of the synergistic interaction of contact stresses and shear forces at the frictional interface. This finding indicates that, directly within the contact zone, there is a simultaneous initiation of a phase transformation of carbon structures with sp^2 bond hybridization to sp^3 hybridization. This transformation is initiated upon a very short time of interaction between the coating surface and the counterbody, and occurs in conjunction with progressive graphitization processes. However, as this transformation is metastable, between successive friction cycles, the newly formed sp^3 structures undergo a transformation back to sp^2 bond hybridization [38, 44]. Figure 4.10 illustrates a schematic representation of the metastable transformation of the chemical structure of carbon coatings as a consequence of stresses and shear forces in the contact zone.

4.4 The Role of Third-Body Interactions in Tribology of Diamond-Like Carbon Coatings

In the previous paragraph, the mechanism of the phase transformation of the coating material induced by the operating parameters of the friction contact was presented. In the immediate vicinity of the contact zone, a transition layer of limited thickness is formed with a graphite-like structure and properties, characterised by low shear resistance. Wear on the surfaces of the mating components is an inherent phenomenon associated with frictional processes. The resulting debris particles are predominantly carbon based, with the addition of other elements from the mating surface. The resulting wear products can remain in the contact zone and continue to participate in the tribological processes taking place. Alternatively, depending on the nature of the mating and the conditions of cooperation, the wear products may be removed from the area of interaction between the frictional pair. If the wear products are retained in the contact zone of the mating surfaces, they undergo significant mechanical fragmentation and may also undergo chemical reactions with the mating material and components of the surrounding atmosphere. It is often observed that these products are transferred from one element of the friction pair to the other. They can also form a continuous layer on the surface of the latter, acting as a solid lubricant. The formation of the transition layer on the surface of the counterbody starts at the beginning of the interaction of the friction pair, during the so-called running-in phase. The layer, which grows gradually, has a low-density amorphous structure consisting of numerous locally ordered nanoclusters composed mainly of sp^2-hybridised carbon atoms, with traces of oxidation products of the counterbody material (in the case of tests carried out under normal conditions). [42]. The results of tests on the formation of the transition layer in controlled working atmospheres show that it is a key component in achieving exceptionally low and stable coefficients of friction. In addition, its superior lubricating properties provide optimal wear protection for both the coating and the counterbody material, while facilitating the local smoothing of their roughness peaks [45]. The kinetics and thermodynamics of transition layer formation are contingent on the type and chemical composition of the counterbody, as well as its surface topography and friction node operating parameters (relative sliding speed and load). Materials characterised by a high chemical affinity for carbon exhibit a greater propensity and speed of transition layer formation, which in such cases takes the character of a continuous and well-adhered coating appearing on the substrate over the entire contact surface [4]. In Fig. 4.11 is presented the SEM of the wear scar on 100Cr6 counterbody and the wear track on DLC coating after a ball-on-disc test. On the surface of the counterbody a continuous and well adherent transfer layer is visible. Moreover, the whole surface of the wear scar is uniformly covered.

In contrast, in the case of materials with a low chemical affinity for carbon, the transfer of wear products to the surface of the counterbody is not as intense, although it is still discernible. As illustrated in Fig. 4.12, a buildup layer of debris particles is

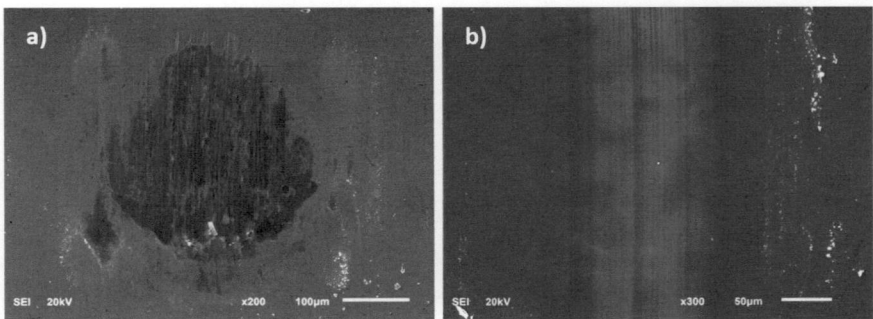

Fig. 4.11 SEM view of the transfer layer on the surface of 100Cr6 countebody and the corresponding wear track on DLC layer after 1000 m ball-on-disc test under load of 10 N in ambient atmosphere and sliding speed of 0.1 m/s. Reprinted with permission from [46]

4.4 The Role of Third-Body Interactions in Tribology of Diamond-Like ...

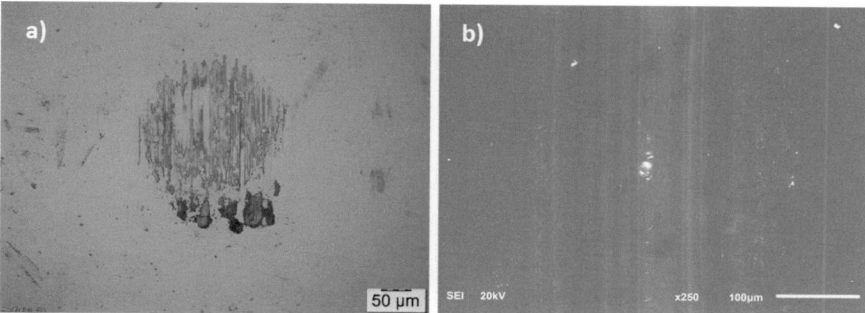

Fig. 4.12 **a** OM image of the transfer layer on the surface of ZrO_2 countebody and **b** SEM image of the corresponding wear track on DLC layer after 1000 m ball-on-disc test under load of 10 N in ambient atmosphere and sliding speed of 0.1 m/s. Reprinted with permission from [46]

mainly visible at the bottom of the wear scar in the direction opposite to the relative movement of the friction node, indicating a hindered process of its formation.

In the case of hard ceramic-based materials, the formation of the transition layer, as well as the maintenance of stable and low friction coefficient values, are dependent on the mechanical nature and adhesive interactions of the friction pair surface. In the case of the Al_2O_3 counterbody, the formation of a transition layer was observed to result in the attainment of low and stable values of the friction coefficient, as well as negligible wear of the mating surfaces. This phenomenon was observed during tests conducted under high vacuum and 2N load, and attributed to the low value of the adhesive interaction forces between Al_2O_3 and the hydrogenated carbon coating. In contrast, the ZrO_2 counterbody exhibited a distinctly different behaviour. Despite an initial low value of the friction coefficient, the operating time of the friction pair in the superlubricity regime was considerably shorter, resulting in pronounced wear of the counterbody material. The results of tests carried out under identical conditions for counterbodies made of SiC and Si_3N_4 demonstrated a divergent character. The superlubricity failure of the Si_3N_4/DLC and SiC/DLC friction pairs was attributed primarily to severe surface damage to the DLC films. This occurred notwithstanding the maintenance of stable transfer layers on the ball surfaces throughout the entirety of the sliding processes. The primary distinction between the wear character of ZrO_2, SiC and Si_3N_4 was the lower hardness and modulus of elasticity of the ZrO_2 counterbody, coupled with robust adhesive interactions between the mating surfaces, which led to progressive wear of the ZrO_2 surface. An increase in the contact area resulted in a reduction in the value of contact stresses, which in turn led to a decrease in the intensification of graphitisation and transition layer formation processes. In the case of SiC and Si_3N_4 counterbodies high interfacial adhesion strength in conjunction with their comparatively superior mechanical properties in relation to the DLC coating, resulted in the severe wear of the coating [47]. The corresponding SEM and Raman spectra of the analysed DLC surfaces are depicted in Fig. 4.13, whereas topography and the cross-section profiles of the coutnerbodies with corresponding Hertz contact pressures are presented in Fig. 4.14.

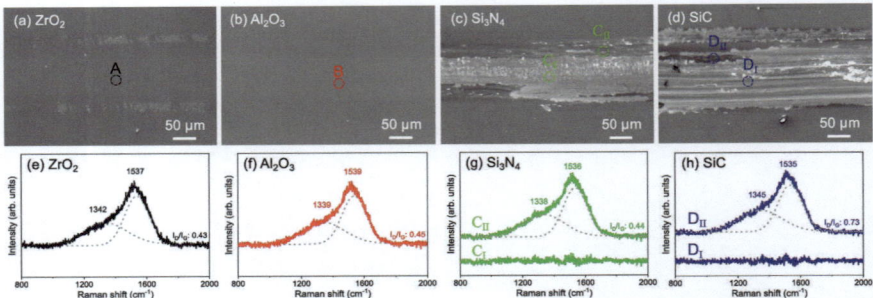

Fig. 4.13 SEM and Raman characterizations of the worn DLC surfaces sliding against four counterbodies: **a** and **e** ZrO_2, **b** and **f** Al_2O_3, **c** and **g** Si_3N_4, **d** and **h** SiC. Reprinted with permission from [47]

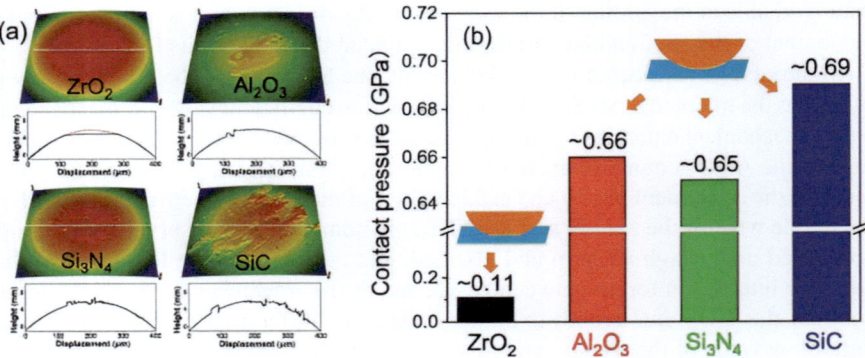

Fig. 4.14 a Topography and the corresponding cross-section profiles of the used counterbodies. **b** corresponding Hertz contact pressures between DLC film and the four used counterbodies. Reprinted with permission from [47]

It is imperative to acknowledge that the conditions under which the test is conducted exerts a substantial influence on its outcome, with the atmosphere and the forces acting upon the friction pair potentially resulting in interactions of a divergent nature between the mating surfaces. Figure 4.15 presents an image and cross-section profile of the wear track of the carbon coating in conjunction with the ZrO_2 counterbody when the friction node is subjected to a force of 20 N under ambient conditions. The creation of a transition layer has been shown to result in localised smoothing of the roughness asperities without significant volume loss, as well as negligible wear on the surface of the ball material [12].

The formation of a transition layer on the mating surface of ceramic-DLC friction contacts is a key element in achieving low and stable friction coefficient values while maintaining high wear resistance. However, it seems reasonable to search for other factors that may influence this process. The Al_2O_3-DLC friction pairs have been shown to exhibit excellent tribological properties in other literature reports;

4.4 The Role of Third-Body Interactions in Tribology of Diamond-Like …

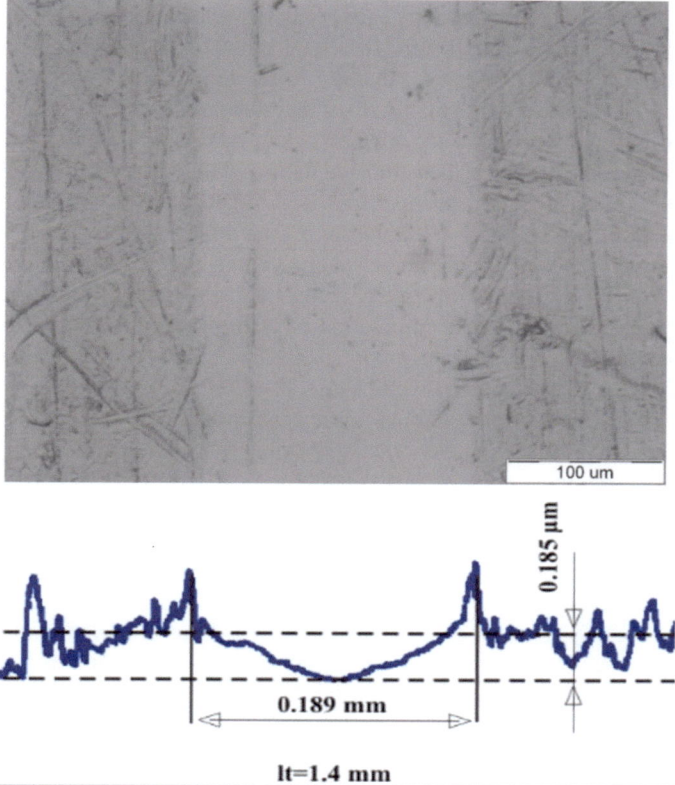

Fig. 4.15 Cross section profile and image of the wear track of the carbon coating cooperating with the ZrO$_2$ counterbody under ambient conditions and 20 N load. Reprinted with permission under Creative Commons license form [12]

however, the authors explicitly point out the dependence of the transition layer formation kinetics on the relative sliding speed of the friction node. The hindered formation of a continuous and well bonded transition layer is identified with insufficient interaction of the friction pair elements in the contact zone and vibrations caused by high rotational speed [48].

Notwithstanding the considerable influence of surface roughness on the tribological properties of DLC coatings, this factor is frequently overlooked in testing, with the majority of studies conducted on polished substrates with a mirror finish. Conversely, the production of polished surfaces is a time-consuming and labour-intensive process, rendering it infeasible for components with complex shapes in the majority of applications. It is therefore necessary to determine whether it is possible to reach a compromise between the quality of the surface finish and the tribological properties of DLC coatings in terms of forming a transition layer on the surface of the counterbody. Research has shown that there is not always a need to prepare a mirror-polished surface finish. Instead, highly accurate treatment on high-graded abrasive papers can give satisfactory results in terms of lowering the friction coefficient and ensuring high wear resistance [49]. It is imperative to acknowledge that the formation process will be directly disrupted during the running-in period, when a transition layer is formed on the surface of the counterbody, due to the abrasive effect of the roughness peaks. The abrasive nature of this interaction will also result in the formation of considerable quantities of wear products derived from both the coating itself and the counterbody material. Concentrations of oxidised wear products resulting from this process can act as abrasive agents, hindering the formation of a stable transition layer. Conversely, a carefully designed geometrical structure of the surface can lead to the accumulation of wear products in the recesses of the profile. In certain instances, this can contribute to a reduction in friction coefficient values in comparison to polished substrates [50]. Figure 4.16 provides a schematic representation of the kinetics of transition layer formation on the surface of a steel counterbody.

In conclusion it can be stated that the tribological properties of carbon coatings are dependent on a number of factors, both internal and external. The aforementioned phenomenological descriptions of individual phenomena, in addition to interpretations of their kinetics of formation and subsequent development, provide a qualitative basis for considering the broad spectrum of tribological properties of carbon coatings. However, this is not a general overview, but rather an examination of each phenomenon individually. The resulting friction force is the result of the interactions

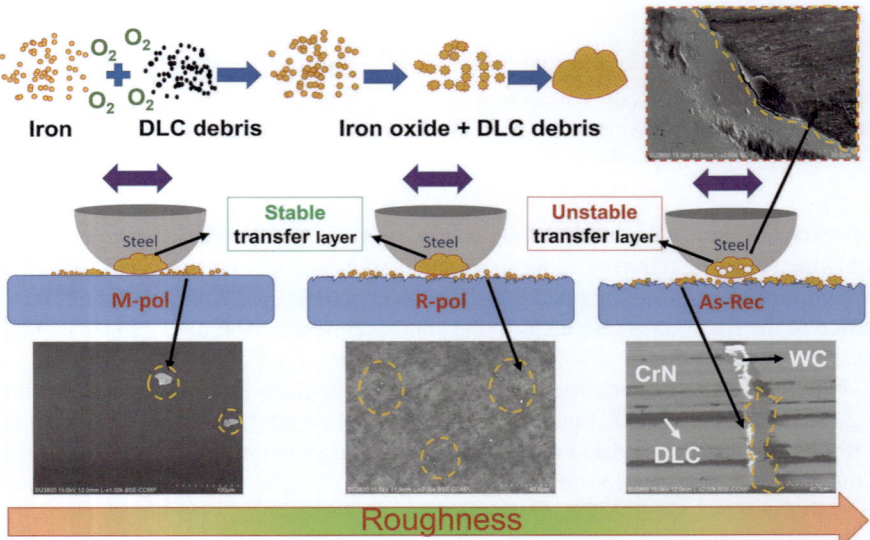

Fig. 4.16 Schematic showing the formation kinematics of transfer layer on steel counterbody sliding against DLC coated WC substrate having different surface roughness. Reprinted with permission under Creative Commons CC BY from [50]

between mechanical surface imperfections, passivation and changes in the chemical structure of the coatings, as well as the formation and subsequent fate of friction pair wear products. The mutual contribution of these factors to the development of specific tribological properties in a quantitative manner is challenging to describe, due to the influence of the type of coating, the counterbody material, the loading conditions, the kinetics and the operating atmosphere of the friction contact.

References

1. Hauert, R.: An overview on the tribological behavior of diamond-like carbon in technical and medical applications. Tribol. Int. **37**, 991–1003 (2004). https://doi.org/10.1016/j.triboint.2004.07.017
2. Hauert, R., Muller, U.: An overview on tailored tribological and biological behavior of diamond-like carbon. Diam. Relat. Mater. **12**, 171–177 (2003)
3. Erdemir, A.: Genesis of superlow friction and wear in diamondlike carbon films. Tribol. Int. **37**, 1005–1012 (2004). https://doi.org/10.1016/j.triboint.2004.07.018
4. Erdemir, A., Donnet, C.: Tribology of diamond-like carbon films: recent progress and future prospects. J. Phys. D Appl. Phys. **39** (2006). https://doi.org/10.1088/0022-3727/39/18/R01
5. Donnet, C., Erdemir, A. (eds.): Tribology of Diamond-Like Carbon Films. Springer, US, Boston, MA (2008)
6. Holmberg, K., Matthews, A.: Coatings Tribology: Properties Techniques and Applications in Surface Engineering. Elsevier, Amsterdam (1994)
7. Matthews, A., Franklin, S., Holmberg, K.: Tribological coatings: contact mechanisms and selection. J. Phys. D Appl. Phys. **40**, 5463–5475 (2007). https://doi.org/10.1088/0022-3727/40/18/S07
8. Holmberg, K., Matthews, A.: Tribological properties of metallic and ceramic coatings. In: Bhushan, B. (ed.) Modern Tribology Handbook. CRC Press (2000)
9. Holmberg, K., Matthews, A.: Coatings Tribology—Properties, Mechanisms, Techniques and Applications in Surface Engineering. Elsevier (2009)
10. Vetter, J.: 60 years of DLC coatings: historical highlights and technical review of cathodic arc processes to synthesize various DLC types, and their evolution for industrial applications. Surf. Coat. Technol. **257**, 213–240 (2014). https://doi.org/10.1016/j.surfcoat.2014.08.017
11. Peng, Y., Peng, J., Wang, Z., Xiao, Y., Qiu, X.: Diamond-like carbon coatings in the biomedical field: properties. Appl. Future Dev. Coat. **12**, 1088 (2022). https://doi.org/10.3390/coatings12081088
12. Batory, D., Szymanski, W., Clapa, M.: Mechanical and tribological properties of gradient a-C:H/Ti coatings. Mater. Sci. Poland **31**, 415–423 (2013). https://doi.org/10.2478/s13536-013-0121-9
13. Sedlaček, M., Vilhena, L.M.S., Podgornik, B., Vižintin, J.: Surface topography modelling for reduced friction. Strojniški vestnik—J. Mech. Eng. **57**, 674–680 (2011). https://doi.org/10.5545/sv-jme.2010.140
14. Dai, W., Shi, Y., Wang, Q., Wang, J.: Smooth diamond-like carbon films prepared by cathodic vacuum arc deposition with large glancing angles. Diam. Relat. Mater. **141**, 110672 (2024). https://doi.org/10.1016/j.diamond.2023.110672
15. Erdemir, A., Eryilmaz, O.: Achieving superlubricity in DLC films by controlling bulk, surface, and tribochemistry. Friction **2**, 140–155 (2014). https://doi.org/10.1007/s40544-014-0055-1
16. Kwok, S.C.H., Zhang, W., Wan, G.J., McKenzie, D.R., Bilek, M.M.M., Chu, P.K.: Hemocompatibility and anti-bacterial properties of silver doped diamond-like carbon prepared by pulsed filtered cathodic vacuum arc deposition. Diam. Relat. Mater. **16**, 1353–1360 (2007). https://doi.org/10.1016/j.diamond.2006.11.001

17. Takikawa, H., Izumi, K., Miyano, R., Sakakibara, T.: DLC thin film preparation by cathodic arc deposition with a super droplet-free system. Surf. Coat. Technol. **163–164**, 368–373 (2003). https://doi.org/10.1016/S0257-8972(02)00629-1
18. Bowden, F.P., Tabor, D.: Friction and Lubrication of Solids. The Clarendon Press, Oxford (1964)
19. Kluba, A., Bociaga, D., Dudek, M.: Hydrogenated amorphous carbon films deposited on 316L stainless steel. Diam. Relat. Mater. **19**, 533–536 (2010). https://doi.org/10.1016/j.diamond.2009.12.020
20. Erdemir, A., Nilufer, I.B., Eryilmaz, O.L., Beschliesser, M., Fenske, G.R.: Friction and wear performance of diamond-like carbon films grown in various source gas plasmas (1999)
21. Fontaine, J., Donnet, C., Grill, A., Lemogne, T.: Tribochemistry between hydrogen and diamond-like carbon films. Surf. Coat Technol. 286–291 (2001)
22. Donnet, C., Erdemir, A.: Historical developments and new trends in tribological and solid lubricant coatings. Surf. Coat. Technol. **180–181**, 76–84 (2004). https://doi.org/10.1016/j.surfcoat.2003.10.022
23. Donnet, C., Grill, A.: Friction control of diamond-like carbon coatings (1997)
24. Li, H., Xu, T., Wang, C., Chen, J., Zhou, H., Liu, H.: Tribochemical effects on the friction and wear behaviors of a-C:H and a-C films in different environment. Tribol. Int. **40**, 132–138 (2007). https://doi.org/10.1016/j.triboint.2006.03.007
25. Andersson, J., Erck, R.A., Erdemir, A.: Friction of diamond-like carbon films in different atmospheres. Wear **254**, 1070–1075 (2003). https://doi.org/10.1016/S0043-1648(03)00336-3
26. Andersson, J., Erck, R.A., Erdemir, A.: Frictional behavior of diamondlike carbon films in vacuum and under varying water vapor pressure. Surf. Coat. Technol. **163–164**, 535–540 (2003). https://doi.org/10.1016/S0257-8972(02)00617-5
27. Gharam, A.A., Lukitsch, M.J., Qi, Y., Alpas, A.T.: Role of oxygen and humidity on the tribochemical behaviour of non-hydrogenated diamond-like carbon coatings. Wear **271**, 2157–2163 (2011). https://doi.org/10.1016/j.wear.2010.12.083
28. Liu, E., Ding, Y.F., Li, L., Blanpain, B., Celis, J.-P.: Influence of humidity on the friction of diamond and diamond-like carbon materials. Tribol. Int. **40**, 216–219 (2007). https://doi.org/10.1016/j.triboint.2005.09.012
29. Li, H., Xu, T., Wang, C., Chen, J., Zhou, H., Liu, H.: Humidity dependence on the friction and wear behavior of diamond-like carbon film in air and nitrogen environments. Diam. Relat. Mater. **15**, 1585–1592 (2006). https://doi.org/10.1016/j.diamond.2005.12.048
30. Al Mahmud, K.A.H., Kalam, M.A., Masjuki, H.H., Mobarak, H.M., Zulkifli, N.W.M.: An updated overview of diamond-like carbon coating in tribology. Crit. Rev. Solid State Mater. Sci. **40**, 90–118 (2015). https://doi.org/10.1080/10408436.2014.940441
31. Kim, H.I., Lince, J.R., Eryilmaz, O.L., Erdemir, A.: Environmental effects on the friction of hydrogenated DLC films. Tribol. Lett. **21**, 51–56 (2006). https://doi.org/10.1007/s11249-005-9008-1
32. Dickrell, P.L., Sawyer, W.G., Erdemir, A.: Fractional coverage model for the adsorption and removal of gas species and application to superlow friction diamond-like carbon. J. Tribol. **126**, 615–619 (2004). https://doi.org/10.1115/1.1739408
33. Cui, L., Lu, Z., Wang, L.: Probing the low-friction mechanism of diamond-like carbon by varying of sliding velocity and vacuum pressure. Carbon N Y **66**, 259–266 (2014). https://doi.org/10.1016/j.carbon.2013.08.065
34. Matta, C., Eryilmaz, O.L., De Barros Bouchet, M.I., Erdemir, A., Martin, J.M., Nakayama, K.: On the possible role of tribopasma in friction and wear of diamond-like carbon films in hydrogen-containing environments. J. Phys. D Appl. Phys. **42** (2009). https://doi.org/10.1088/0022-3727/42/7/075307
35. Nakayama, K.: Triboemission of electrons, ions, and photons from diamondlike carbon films and generation of tribomicroplasma. Surf. Coat. Technol. **188–189**, 599–604 (2004). https://doi.org/10.1016/j.surfcoat.2004.07.103
36. Ni, W., Cheng, Y.T., Weiner, A.M., Perry, T.A.: Tribological behavior of diamond-like-carbon (DLC) coatings against aluminum alloys at elevated temperatures. Surf. Coat. Technol. **201**, 3229–3234 (2006). https://doi.org/10.1016/j.surfcoat.2006.06.045

37. Kim, D.-W., Kim, K.-W.: Effects of sliding velocity and normal load on friction and wear characteristics of multi-layered diamond-like carbon (DLC) coating prepared by reactive sputtering. Wear **297**, 722–730 (2013). https://doi.org/10.1016/j.wear.2012.10.009
38. Zhou, Y., Chen, Z., Zhang, T., Zhang, S., Xing, X., Yang, Q., Li, D.: Metastable hybridized structure transformation in amorphous carbon films during friction—a study combining experiments and MD simulation. Friction **11**, 1708–1723 (2023). https://doi.org/10.1007/s40544-022-0690-x
39. Scharf, T.W., Prasad, S.V.: Solid lubricants: a review. J. Mater. Sci. **48**, 511–531 (2013). https://doi.org/10.1007/s10853-012-7038-2
40. Scharf, T.W., Ohlhausen, J.A., Tallant, D.R., Prasad, S.V.: Mechanisms of friction in diamondlike nanocomposite coatings. J. Appl. Phys. **101** (2007). https://doi.org/10.1063/1.2711147
41. Singer, I.L., Bolster, R.N., Wegand, J., Fayeulle, S., Stupp, B.C.: Hertzian stress contribution to low friction behavior of thin MoS_2 coatings. Appl. Phys. Lett. **57**, 995–997 (1990). https://doi.org/10.1063/1.104276
42. Chen, X., Zhang, C., Kato, T., Yang, X.A., Wu, S., Wang, R., Nosaka, M., Luo, J.: Evolution of tribo-induced interfacial nanostructures governing superlubricity in a-C:H and a-C:H:Si films. Nat. Commun. **8** (2017). https://doi.org/10.1038/s41467-017-01717-8
43. Ma, T.-B., Hu, Y.-Z., Wang, H.: Molecular dynamics simulation of shear-induced graphitization of amorphous carbon films. Carbon N Y **47**, 1953–1957 (2009). https://doi.org/10.1016/j.carbon.2009.03.040
44. Chen, Z., Xing, X., Zhang, T., Zhang, S., Yang, Q., Zhang, B., Gao, K., Zhou, Y.: Friction-induced metastable transformation of amorphous carbon film: Exploration by experimental and molecular dynamics simulations. Appl. Surf. Sci. **628** (2023). https://doi.org/10.1016/j.apsusc.2023.157327
45. Yin, X., Mu, L., Jia, Z., Pang, H., Chai, C., Liu, H., Liang, C., Zhang, B., Liu, D.: Nanostructure of superlubricating tribofilm based on friction-induced a-C: H films under various working conditions: a review of solid lubrication. Lubricants **12**, 40 (2024). https://doi.org/10.3390/lubricants12020040
46. Jedrzejczak, A., Kolodziejczyk, L., Szymanski, W., Piwonski, I., Cichomski, M., Kisielewska, A., Dudek, M., Batory, D.: Friction and wear of a-C:H:SiOx coatings in combination with AISI 316L and ZrO_2 counterbodies. Tribol. Int. **112**, 155–162 (2017). https://doi.org/10.1016/j.triboint.2017.03.026
47. Liu, Y., Jiang, Y., Sun, J., Wang, L., Liu, Y., Chen, L., Zhang, B., Qian, L.: Durable superlubricity of hydrogenated diamond-like carbon film against different friction pairs depending on their interfacial interaction. Appl. Surf. Sci. **560**, 150023 (2021). https://doi.org/10.1016/j.apsusc.2021.150023
48. Liu, Y., Yu, B., Cao, Z., Shi, P., Zhou, N., Zhang, B., Zhang, J., Qian, L.: Probing superlubricity stability of hydrogenated diamond-like carbon film by varying sliding velocity. Appl. Surf. Sci. **439**, 976–982 (2018). https://doi.org/10.1016/j.apsusc.2018.01.048
49. Soprano, P.B., Salvaro, D.B., Giacomelli, R.O., Binder, C., Klein, A.N., de Mello, J.D.B.: Effect of soft substrate topography on tribological behavior of multifunctional DLC coatings. J. Braz. Soc. Mech. Sci. Eng. **40** (2018). https://doi.org/10.1007/s40430-018-1290-6
50. Khan, S.A., Oliveira, J., Ferreira, F., Emami, N., Ramalho, A.: Surface roughness influence on tribological behavior of HiPIMS DLC coatings. Tribol. Trans. **66**, 565–575 (2023). https://doi.org/10.1080/10402004.2023.2197472

Chapter 5
Doped Diamond-Like Carbon Coatings with Metallic and Nonmetallic Elements

Abstract The chapter examines the modification of the chemical composition of carbon coatings with a view to enhancing their tribological properties. The impact of the most prevalent dopants on reducing the coefficient of friction and enhancing wear resistance, while ensuring good coating adhesion to the substrate material, was analysed. Specific attention is devoted to elements such as fluorine, silicon and both carbide-forming and non-carbide-forming metals.

In the previous chapter it has been shown that carbon coatings represent the optimal solution for a multitude of applications, encompassing all potential facets of friction and wear. As is already established, these coatings have acquired a reputation based on their high hardness, excellent resistance to abrasive wear, and very low friction coefficient values, which have been demonstrated in a variety of working atmospheres. However, they also exhibit significant limitations, with the most challenging being the considerable residual stress. When deposited on metallic substrates, the discrepancy between the thermal expansion coefficient of the substrate and that of the DLC coating results in the generation of micro strain within the amorphous carbon matrix, which in turn gives rise to high residual compressive stress. This represents a significant challenge for the fabrication of DLC layers, as the accumulation of high residual stress restricts their thickness and may lead to delamination of the coating. A variety of techniques have been employed to reduce residual stress in DLC coatings. These include the deposition of adhesion-promoting interlayers comprising uniform, multilayered, graded compositions; the application of different surface treatments, such as plasma etching or thermo-chemical treatment of the substrate material; and the modification of the chemical composition of the DLC coating through the incorporation of different non-metallic and metallic elements. It has been demonstrated that these approaches result in noticeable changes in the properties of the coatings, including their tribological behaviour, which is beneficial in terms of reducing stress and improving adhesion. The topic of improving the adhesion of carbon coatings is a broad and complex, and could constitute a separate literature item in itself. The selection of an appropriate interlayer, multilayer structure and processing parameters is of paramount importance for enhancing adhesion, and encompasses a multitude of technical considerations. While it exerts a degree of influence on the tribological parameters of DLC coatings, the underlying amorphous carbon coating remains unaltered. Consequently, the present discussion will focus on the fundamental principles and recent advancements in the incorporation of diverse elements into DLC matrix and their impact on the resulting tribological parameters.

The doping strategies employed in the incorporation of diverse elements into DLC matrix typically involve the introduction of non-metallic and metallic elements, which either form bonds with carbon atoms or precipitate pure/carbide nanocrystalline inclusions that are uniformly distributed within an amorphous carbon matrix. The most prevalent non-metallic dopants include fluorine, nitrogen and silicon, while silver, titanium, zirconium, tungsten and chromium are among the most commonly considered metallic components [1]. The impact of specific elements on the tribological characteristics of the resulting layers is frequently determined not only by the chemical composition of the coating and its interaction with the counterbody, but also by the overall reduction of residual stress, thereby enhancing adhesion to a range of metallic substrates. Consequently, in this discussion, only those elements that impart additional functionality to the coating will be addressed.

5.1 Fluorinated DLC

Fluorine is a noteworthy admixture material, largely due to its distinctive chemical properties. This highly reactive element is characterised by the highest electronegativity in the periodic table, which precludes it from forming a double bond with carbon. Instead, the resulting C–F bonds contribute to the reduction in length of chains and rings in the DLC structure, leading to a decrease in the amount of sp^3 hybridised carbon bonds and a reduction in the density and compactness of the carbon network. Concurrently, the coating's low residual stress ensures enhanced adhesion to the substrate [1]. The chemical structure of F-DLC coatings has been shown to exhibit a reduced concentration of hydrogenated carbon compounds, with a concomitant increase in the prevalence of CF_2 and CF_3 groups. This phenomenon has been demonstrated to result in a decrease in network density, thereby leading to a reduction in the value of the dispersive component of surface free energy, which approaches the level characteristic of PTFE—18 mNm^{-1} [2]. It is for this reason that fluorinated carbon coatings are often referred to as 'superhard Teflon'. These coatings combine superhydrophobicity and superlubricity properties, with a hardness that is slightly lower than that of the unmodified DLC, but still superior in comparison to PTFE [3]. Furthermore, fluorine-terminated DLC surfaces have been shown to generate higher repulsive electrostatic forces, which contribute to low friction in high humidity atmospheres [4]. Additionally, the ease with which droplets roll off these surfaces provides excellent tribo-corrosion and anti-icing properties, making them suitable for use in aircraft engineering applications [5]. The incorporation of fluorine into the DLC matrix has been shown to reduce the sp^3 hybridised carbon concentration and promote the formation of more sp^2 hybridised carbon domains. Unlike

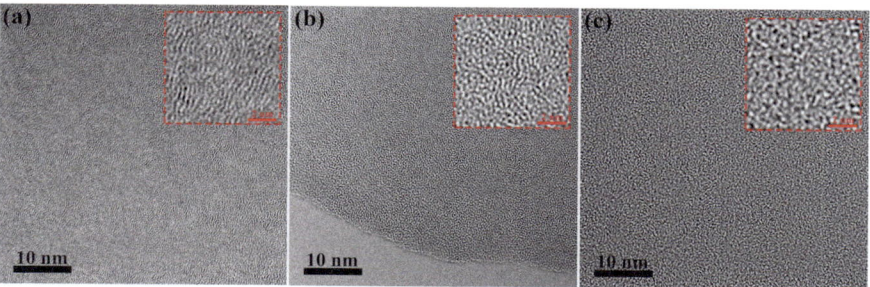

Fig. 5.1 HRTEM plane-view images of F-DLC films with different fluorine contents: 4.8 at.% (**a**), 12.1 at.% (**b**), and 15.5 at.% (**c**). The insets are the corresponding partial enlarged details. Reprinted with permission from [6]

other doping non-metallic elements, fluorine influences the microstructure of the coating. The newly formed C-F bonds break the cross-linked carbon network, which consequently leads to the collapse of the sp^3 hybridised matrix. The microstructure ultimately transforms with increasing dopant concentration. Wang et al. [6, 7] examined the impact of varying levels of fluorine incorporation into diamond-like carbon on the resultant microstructure, mechanical and tribological properties. For coatings with comparatively low fluorine concentrations, a substantial amount of discontinuous nanometer-sized ordered structures, alternately arranged within the amorphous carbon matrix, was reported. These well-ordered ring and fingerprint-shaped domains, composed of curved and straight graphitic planes, were tightly interlocked by C–C crosslink bonds, resembling fullerene-like structures. As the fluorine concentration was increased, the C–F groups accumulated and CF2 groups emerged in the carbon matrix. This resulted in a reduction in the size and an increase in the disorder of the fullerene-like structures, until they completely disappeared, being replaced by polymer-like chains. The coating thus presents a typical amorphous structure. The microstructure changes depending on the fluorine content, are illustrated in Fig. 5.1.

The mechanical and tribological properties of fluorinated carbon coatings are predominantly influenced by the concentration of fluorine. The extant literature consistently highlights that low concentrations of fluorine result in a reduction in residual stress and enhancement of the adhesion of the coating. However, as the concentration of the dopant increases, there is a gradual decline in hardness and wear resistance [7, 8]. However, some authors, have identified changes in mechanical properties with the appearance of fullerene-like structures, suggesting a dual impact of fluorine on DLC properties. Coatings with low concentrations of F and nanometer-size ordered fullerene-like structures exhibit an excellent elastic recovery [6]. The reported values of the coefficient of friction have been observed to oscillate below $\mu = 0.06$ in dry environment [1] and between 0.06 and 0.08 in physiological saline solution [7].

5.2 Silicon Incorporated Carbon Coatings

In the field of diamond-like carbon (DLC) coating modification, silicon stands out as a particularly noteworthy element. Regardless of the properties under scrutiny or the intended application, silicon-doped carbon coatings have exhibited superior performance in all aspects when compared to unmodified DLC. These include high corrosion resistance [9, 10], high thermal stability [11], and very good biological properties [12, 13]. Consequently, it can be deduced that silicon doping significantly enhances the mechanical and tribological properties of carbon coatings. A notable benefit of silicon doping is the ability to reduce residual stress. The primary mechanism by which silicon exerts its effect on this parameter is believed to be the binding of free hydrogen atoms that are not chemically bonded to the coating, thereby occupying its voids [14]. The resultant reduction in residual compressive stress value ensures enhanced adhesion and increased coating thickness without risk of delamination. Doping with silicon generally results in an increased ID/IG parameter; however, this is predominantly related to the tetravalent structure of silicon and its four-fold coordinated network, which is characteristic of Si–C or diamond [10]. The hardness of DLC coatings is inextricably linked to the content of sp^3 hybridized carbon bonds, yet some authors have reported increased hardness and decreased residual stress of silicon incorporated carbon coatings deposited using a silane/methane working atmosphere [15]. In every other case the beneficial influence of silicon on the registered coating performance was obtained at the cost of a slight reduction in hardness. Furthermore, Si-containing carbon films are recognised for their exceptionally low friction coefficient. The positive influence of Si on the tribological properties is predominantly evident in the formation of Si–H or Si–OH functional groups on the surface, which can result in a reduction of friction force [16]. However, it should be noted that the presence of silicon has been shown to be accompanied by a significant increase in wear of the coating, particularly under dry friction conditions [17]. Nevertheless, the results of friction tests conducted at elevated temperatures have demonstrated an opposing trend, with silicon exhibiting a positive influence on high-temperature wear and oxidation resistance up to 450 °C. The stabilisation of sp^3 bonded carbon structures (preventing excessive graphitisation of the coating) and the formation of Si–O–C bonds in the wear track (ensuring an additional lubrication effect during sliding) are reported as possible reasons [1, 11, 17, 18].

Batory et al. conducted a thorough investigation into the characteristics of silicon-incorporated a-C:H:SiO$_x$ coatings in terms of surface topography, chemical structure, as well as corrosion and mechanical properties versus applied process parameters [9, 14, 19]. The results revealed a clear dependence of coating properties on both, the concentration of silicon and oxygen, as well as the applied negative bias voltage during the PA CVD deposition process. The coatings deposited under a bias of − 600 V exhibited minimal decline in hardness with increasing silicon concentration, while concurrently demonstrating an enhancement in both H/E and H^3/E^2 ratios (see Fig. 5.2). Conversely, coatings characterised by a defect-free, homogeneous

5.2 Silicon Incorporated Carbon Coatings

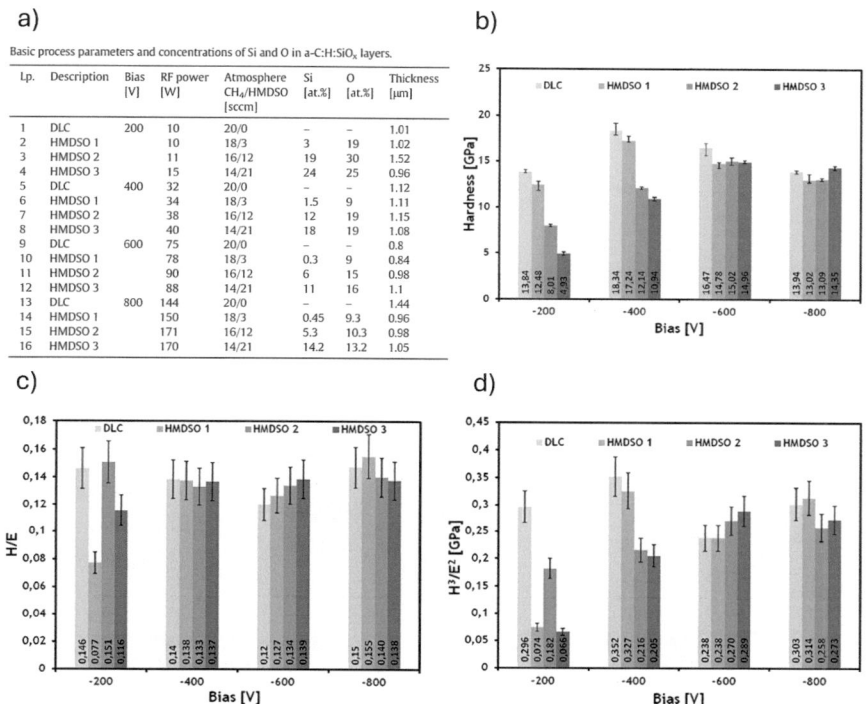

Fig. 5.2 Characteristics of a-C:H: SiO_x coatings deposited using different self-bias voltages and CH_4/HMDSO flow ratios: **a** process parameters and concentrations of Si and O; **b** hardness; **c** H/E ratio and **d** H^3/E^2 ratio. Reprinted with permission from [14]

surface structure were obtained for a bias of −800 V, yielding highly satisfactory and predictable outcomes in corrosion resistance tests.

Jedrzejczak et al. [20, 21] subsequently investigated the same coatings in terms of wear and friction for different counterbody materials and environmental sliding conditions (see Fig. 5.3, which presents the evolution of the coefficient of friction of a-C:H:SiO_x coatings tested under dry sliding conditions against AISI316L and ZrO_2 counterbodies). A substantial positive impact of SiO_x on the reduction of the friction coefficient was observed for coatings tested with AISI316L counterbody. In the case of zirconia, the high chemical affinity of SiO_x to ZrO_2 resulted in a notable increase in the value of the coefficient of friction and a general deterioration of its course with increasing silicon concentration. However, as the authors state, even a low concentration of silicon and oxygen (0.5 and 9 at.%, respectively) can significantly improve the tribological properties of DLC coatings, regardless of the type of counterbody, when considering their other superior properties. These results, and other reports, also clearly indicate the important role of substrate bias during deposition of a-C:H:SiO_x coatings that eventually determines the tribological behaviour

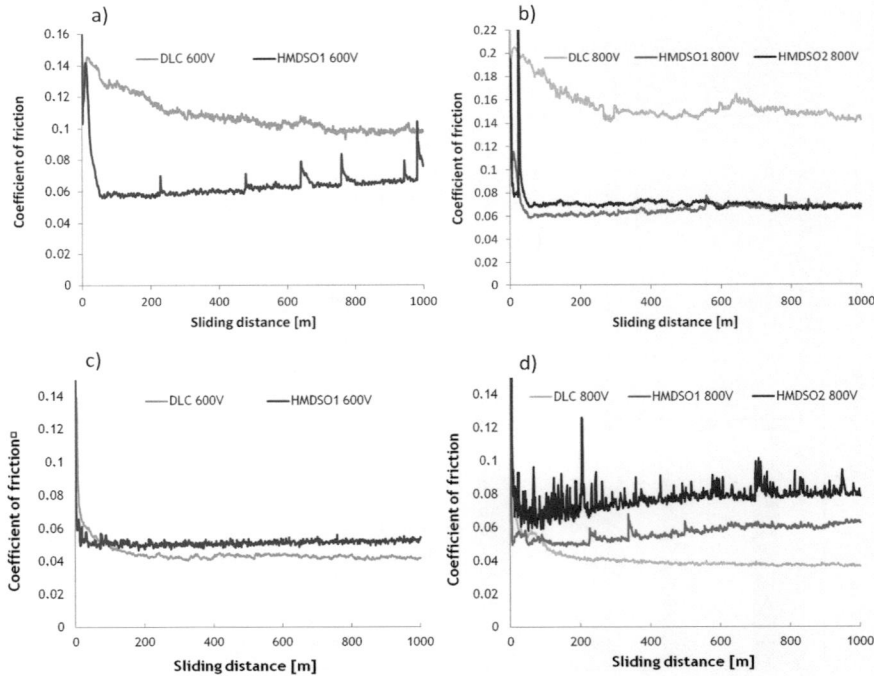

Fig. 5.3 Friction coefficient evolution of DLC and a-C:H:SiO$_x$ coatings synthesized using negative self-bias of 600 and 800 V against: **a** and **b** AISI316L counterbody; **c** and **d** ZrO$_2$ counterbody. Reprinted with permission form [20]

[16]. In terms of the deterioration of anti-wear properties of coatings, the authors also achieved high compliance with other literature reports, registering at least one order of magnitude higher wear for the coatings doped with silicon and oxygen.

Under lubricated sliding conditions, particularly in simulated body fluid (SBF) and bovine serum albumin (BSA), coatings containing silicon and oxygen (HMDSO) have been shown to exhibit significantly superior friction and anti-wear properties in comparison to those comprising solely silicon (TMS). The Fig. 5.4 presents a comparison of the friction coefficient of silicon-incorporated DLC coatings in various environments: air, SBF, and BSA. In the SBF environment, the HMDSO coatings exhibited a reduced friction and wear rate compared to the oxygen-free layers. In the BSA environment, the TMS layers demonstrated superior performance, exhibiting a lower friction coefficient, likely attributable to a reduced protein attachment to the surface. Overall, both the Si-DLC and SiO$_x$-DLC coatings exhibited a lower wear rate compared to DLC in both SBF and BSA. Furthermore, in the case of coatings with the lowest silicon and oxygen content, the counterbody wear was the lowest of all friction pairs tested.

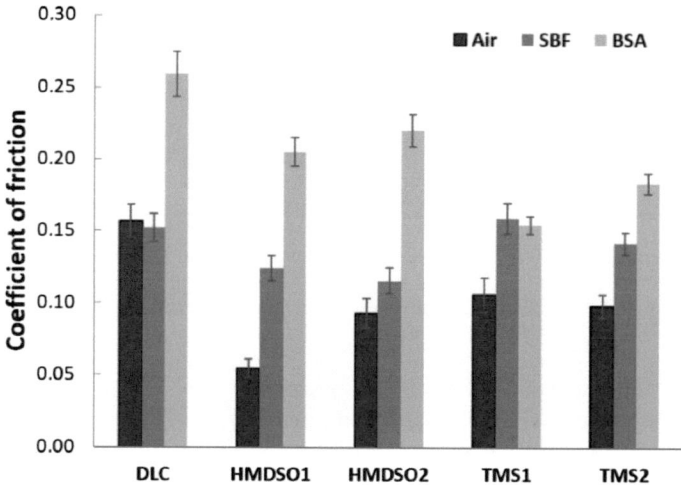

Fig. 5.4 Comparison of the friction coefficient of silicon-incorporated DLC coatings in different environments: air, SBF, and BSA. Reprinted with permission under Creative Commons CC BY from form [21]

As demonstrated by Kołodziejczyk et al. [22] the coefficient of friction and wear rate of silicon-incorporated carbon coatings tested within the nano-Newton load range exhibit an increase with increasing Si concentration. This finding suggests that the enhanced friction and wear is predominantly attributable to elevated contact pressures and the resultant tribo-induced transformations of the amorphous carbon matrix.

5.3 Metal Incorporated DLC

The integration of metal atoms into amorphous carbon coatings has been shown to result in a substantial alteration of their mechanical properties, with a concomitant significant enhancement of their tribological characteristics. During the deposition of metal-incorporated DLCs, the chemical interaction of the dopant material with carbon can occur in a variety of ways. In the simplest case, metal atoms can be dissolved in carbon solid solution or form pure nanocrystalline metallic inclusions (e.g. Ag, Al, Cu), the prevalence of which depends to a large extent on the concentration of the dopant [23]. The incorporation of non-carbide-forming metals into a carbon matrix has been demonstrated to reduce residual stress and enhance the adhesion of the coating. This reduction in residual stress may be attributable to the

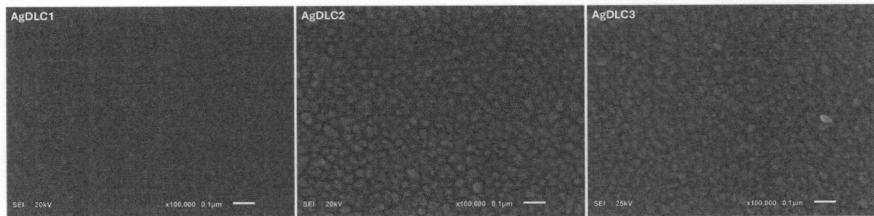

Fig. 5.5 SEM images of silver incorporated DLC coatings: AgDLC1 4.5 at% of Ag, AgDLC2 8.4 at.% of Ag and AgDLC3 15.2 at.% of Ag. The size of Ag coherent diffracting domains determined by XRD was around 12 ± 2 nm. Reprinted with permission form [22]

significantly lower elastic modulus of self-assembled metallic conglomerates, which absorb compressive stress from the carbon matrix [24]. Figure 5.5 presents SEM images of silver-incorporated carbon coatings with different concentrations of Ag.

It has been demonstrated that a significant reduction in the coefficient of friction of Ag-DLC coatings can be achieved through the appropriate selection of the Ag concentration [25]. The authors identified a significant improvement in tribological properties with the low shear strength of Ag clusters on the surface and plastic flow during relative slip. This is attributed to the reduced bonding ability between the nanocrystalline and amorphous phases, which enables their multiple shifts. This, in turn, may also increase the ductility of the film [26]. However, other reports indicate that for higher dopant concentrations, the aggregation of Ag in the wear track and the formation of a silver-rich transfer layer on the counterbody result in a reduction in wear resistance both, at high and low contact pressure [22, 27]. Although the internal stress of the films can be reduced and their tribological properties can be enhanced, these usually come at the cost of a significant decrease in hardness [28]. Nevertheless, aside from tribological functional characteristics, other promising properties for industrial applications exist, including alterations in surface free energy of biocompatibility combined with antibacterial properties [29, 30].

In the context of transition metals, the deposition process gives rise to the formation of metal carbides, which are distributed randomly within an amorphous carbon matrix. This results in the constitution of a novel generation of nanocomposite coatings, which are more commonly referred to as nc–MeC/a-C, where 'nc' denotes 'nanocrystalline' and 'MeC' denotes 'metal carbide'. Owing to their nanocomposite structure, these coatings frequently exhibit properties that are not attainable for unmodified carbon coatings. The size and volume fraction of the nanocrystalline component in the amorphous carbon structure can be readily controlled by varying the deposition parameters, thus allowing for the tailoring of the properties of the nanocomposite coatings. Among the range of materials used, Zr, Ti, W and Cr are by far the most commonly used dopant [1, 31].

5.3 Metal Incorporated DLC

The incorporation of titanium, tungsten or chromium into DLC has been demonstrated to have a significant impact on the mechanical properties and the formation of nanocrystalline size carbides. The shift in carbon hybridisation from sp^3 to sp^2, resulting in a greater proportion of carbide formation, has been shown to lead to a reduction in residual stress and an enhancement in coating adhesion. The nanocrystalline structure of the coating is believed to act as a barrier to micro crack propagation, thereby contributing to enhanced toughness and improved tribological properties. Nevertheless, the most critical issue is the necessity to identify a correlation between the types or amounts of the incorporated elements and the tribological properties of the obtained coating [1, 32]. The Ti-DLC films with micro Ti doping content (1.82 wt%) have been shown to exhibit superior performance in comparison to those of pure DLC films and Ti-DLC films with a relatively high Ti doping content. The presence of TiC nanocrystallites within an amorphous carbon matrix was found to have a significant impact on the mechanical properties of the resultant coatings. It was observed that the size and quantity of the nanocrystallites within the matrix were pivotal in enhancing the hardness of the coatings, as well as improving their tribological properties [33]. Figure 5.6 presents an image from high-resolution transmission electron microscope (HR TEM) of a nanocomposite TiC-DLC coating, illustrating nanometer-sized TiC inclusions separated by an amorphous carbon matrix.

In a separate report on titanium-doped hydrogenated DLC, the authors posit that following the incorporation of titanium and oxygen into a carbon matrix, the resultant coatings exhibited superior friction performance. The coatings were characterised by ultralow and stable friction coefficient (approximately 0.008) in ambient air, minimal sensitivity to relative humidity, irrespective of the counterbody material and the test atmosphere. Interestingly, the structure of the Ti-DLC films revealed lack of

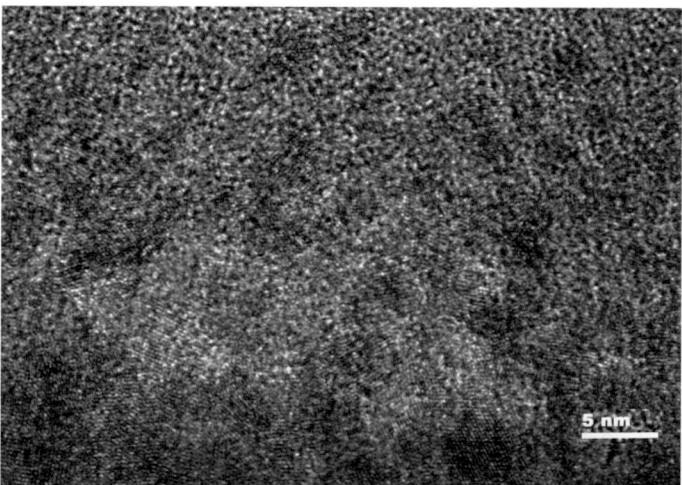

Fig. 5.6 HR TEM view of nanocomposite TiC-DLC coating with nanometer size titanium carbide precipitations embedded in DLC matrix

crystalline TiC inclusions indicating that Ti and O may exist in a form of small size atomic clusters dissolved in amorphous carbon matrix [34].

The doping of carbon coatings with Zr did not result in significant changes to their mechanical properties. However, the friction test results for DLC and Zr-DLC coatings tested at room temperature, 100 °C and in a nitrogen atmosphere clearly demonstrated that the most promising tribological behaviour was achieved for Zr-DLC-H coatings at all tested conditions. The authors posit that a dense and homogeneous low-shear carbonaceous tribolayer is the primary factor in achieving a low friction coefficient and protecting the surface of the counterbody from oxidation processes. The use of Zr in the doping of DLC has been demonstrated to reduce surface free energy and enhance the capacity for the formation of a stable transition layer on the surface of the counterbody. Consequently, the steady-state friction process was attained with considerable rapidity, and the formation of the tribolayer was significantly accelerated. Additionally, the inhibition of the oxidation process of the counterbody led to a substantial reduction in iron oxides, which could have otherwise led to accelerated wear of the friction couple [35, 36].

Tungsten-doped carbon coatings containing nano-scale WC particles, dispersed within an amorphous matrix, have been demonstrated to exhibit excellent tribological properties at elevated temperatures sliding against Ti-6Al-4V counterbody. The low coefficient of friction exhibited by the coatings at a temperature of 25 °C was attributed to the formation of a carbon-rich transition layer on the surface of the counterbody. A gradual increase in temperature up to the 200–300 °C range resulted in a significant increase in the coefficient of friction, accompanied by a substantial increase in coating wear rate. Notably, no transition layer was observed on the surface of the counterebody within this temperature range. However, at temperatures ranging from 400 to 500 °C, a substantial reduction in the adhesion of the titanium counterbody to the test coating was observed, resulting in a notable decrease in the friction coefficient value. The occurrence of a tungsten trioxide layer on the surface of the W-DLC coating was identified as the underlying cause of the reduced wear rate and friction coefficient [37]. Figure 5.7 illustrates the progression of the wear rate of DLC and W-DLC coatings in relation to the temperature at which the test was conducted.

The simultaneous effect of different doping materials on the overall mechanical and tribological properties of carbon-based layers was reported in [38]. Each of the doping elements was precisely matched to impact a certain drawback of DLC: (i) chromium as a strong carbide former causing an increase in hardness, (ii) aluminum, as a weak carbide former, dissolved in the DLC matrix, effectively releasing the high level of residual stress, (iii) silicon, known for its positive influence on the coefficient of friction, additionally enabled the maintenance of the sp^3 structure, thereby enhancing the thermal stability of the coating up 500 °C. It was hypothesised that the AlCrSi multi-doping process could enhance the comprehensive properties of DLC coatings.

5.3 Metal Incorporated DLC

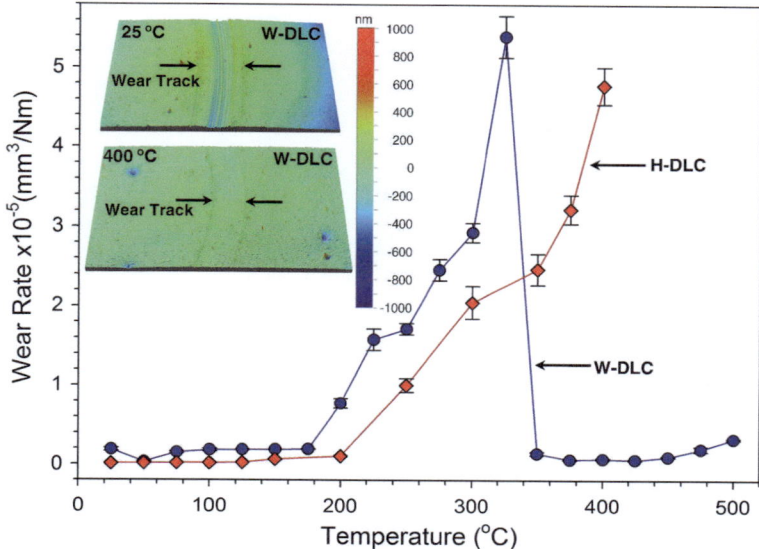

Fig. 5.7 The variations of wear rates (mm³/N-m) of H-DLC and W-DLC against Ti–6Al–4 V with test temperature. Insets show 3-D surface profile images of the wear tracks of W-DLC tested at 25 and 400 °C. Reprinted with permission from [37]

The self-healing properties of tungsten disulfide have led to its use as a dopant for a-C coatings. A pre-notched WS_2/DLC coating was subjected to tribological tests across the previously produced notches under varying loads. At higher frictional contact loads, the notches were completely healed by sliding through rearrangement of the WS_2 wear products forming patchy tribofilm and filling the volume of the notch. In addition, the healed notches acted as excellent micro-reservoirs of lubricant, providing an ultra-low coefficient of friction [39]. An in-situ SEM examination of the partial healing process of the scratch damage is presented in Fig. 5.8.

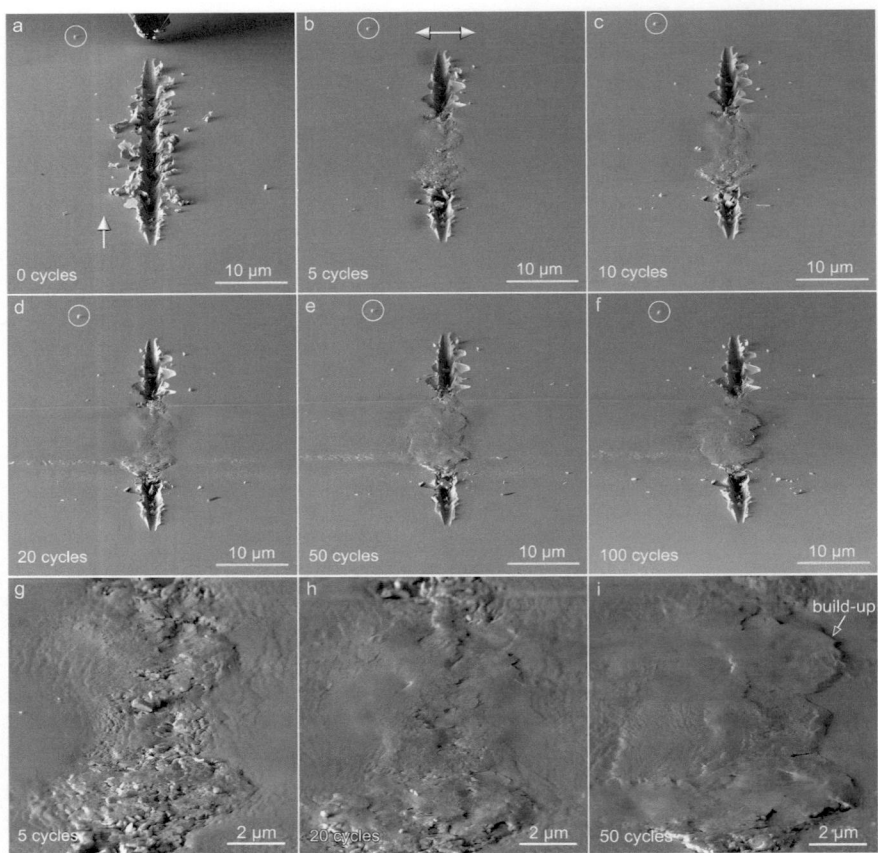

Fig. 5.8 a Initial scratch of WS_2/a-C nanocomposite coating. **b–f** In-situ SEM examination of the partial healing process of the scratch damage by indicated 0–100 reciprocating sliding cycles, under a higher normal load of 500 mN. The arrows in (**a**, **b**) indicate the scratching/reciprocating rubbing directions. **g, h, i** higher magnification images of the healed scratch in (**b, d, e**). The circle in (**a–f**) is an in-situ reference. Reprinted with permission under Creative Commons CC-BY from [39]

References

1. Sharifahmadian, O., Pakseresht, A., Amirtharaj Mosas, K.K., Galusek, D.: Doping effects on the tribological performance of diamond-like carbon coatings: a review. J. Market. Res. **27**, 7748–7765 (2023). https://doi.org/10.1016/j.jmrt.2023.11.132
2. Donnet, C.: Recent progress on the tribology of doped diamond-like and carbon alloy coatings: a review (1998)
3. Sung, J.C., Kan, M.-C., Sung, M.: Fluorinated DLC for tribological applications. Int. J. Refract. Metals Hard Mater. **27**, 421–426 (2009). https://doi.org/10.1016/j.ijrmhm.2008.11.008
4. Bhowmick, S., Sen, F.G., Banerji, A., Alpas, A.T.: Friction and adhesion of fluorine containing hydrophobic hydrogenated diamond-like carbon (F-H-DLC) coating against magnesium alloy AZ91. Surf. Coat. Technol. **267**, 21–31 (2015). https://doi.org/10.1016/j.surfcoat.2014.11.047

References

5. Liu, L., Tang, W., Ruan, Q., Wu, Z., Yang, C., Cui, S., Ma, Z., Fu, R.K.Y., Tian, X., Wang, R., Wu, Z., Chu, P.K.: Robust and durable surperhydrophobic F-DLC coating for anti-icing in aircrafts engineering. Surf. Coat. Technol. **404** (2020). https://doi.org/10.1016/j.surfcoat.2020.126468
6. Wang, J., Zhang, K., Zhang, L., Wang, F., Zhang, J., Zheng, W.: Influence of structure evolution on tribological properties of fluorine-containing diamond-like carbon films: from fullerene-like to amorphous structures. Appl. Surf. Sci. **457**, 388–395 (2018). https://doi.org/10.1016/j.apsusc.2018.06.249
7. Wang, J., Ma, J., Huang, W., Wang, L., He, H., Liu, C.: The investigation of the structures and tribological properties of F-DLC coatings deposited on Ti–6Al–4V alloys. Surf. Coat. Technol. **316**, 22–29 (2017). https://doi.org/10.1016/j.surfcoat.2017.02.065
8. Yao, Z.Q., Yang, P., Huang, N., Sun, H., Wang, J.: Structural, mechanical and hydrophobic properties of fluorine-doped diamond-like carbon films synthesized by plasma immersion ion implantation and deposition (PIII-D). Appl. Surf. Sci. **230**, 172–178 (2004). https://doi.org/10.1016/j.apsusc.2004.02.044
9. Batory, D., Jedrzejczak, A., Kaczorowski, W., Kolodziejczyk, L., Burnat, B.: The effect of Si incorporation on the corrosion resistance of a-C:H:SiOx coatings. Diam. Relat. Mater. **67** (2016). https://doi.org/10.1016/j.diamond.2015.12.002
10. Papakonstantinou, P., Zhao, J.F., Lemoine, P., Mcadams, E.T., Mclaughlin, J.A.: The effects of Si incorporation on the electrochemical and nanomechanical properties of DLC thin films (2002)
11. Choi, J., Nakao, S., Miyagawa, S., Ikeyama, M., Miyagawa, Y.: The effects of Si incorporation on the thermal and tribological properties of DLC films deposited by PBII&D with bipolar pulses. Surf. Coat. Technol. **201**, 8357–8361 (2007). https://doi.org/10.1016/j.surfcoat.2006.02.084
12. Bociaga, D., Kaminska, M., Sobczyk-Guzenda, A., Jastrzebski, K., Swiatek, L., Olejnik, A.: Surface properties and biological behaviour of Si-DLC coatings fabricated by a multi-target DC-RF magnetron sputtering method for medical applications. Diam. Relat. Mater. **67**, 41–50 (2016). https://doi.org/10.1016/j.diamond.2016.01.025
13. Ong, S.E., Zhang, S., Du, H., Too, H.C., Aung, K.N.: Influence of silicon concentration on the haemocompatibility of amorphous carbon. Biomaterials **28**, 4033–4038 (2007). https://doi.org/10.1016/j.biomaterials.2007.05.031
14. Batory, D., Jedrzejczak, A., Szymanski, W., Niedzielski, P., Fijalkowski, M., Louda, P., Kotela, I., Hromadka, M., Musil, J.: Mechanical characterization of a-C:H:SiOx coatings synthesized using radio-frequency plasma-assisted chemical vapor deposition method. Thin Solid Films. **590**, (2015). https://doi.org/10.1016/j.tsf.2015.08.017
15. Damasceno, J., Camargo Jr, S., Freire Jr, F., Carius, R.: Deposition of Si-DLC films with high hardness, low stress and high deposition rates (2000)
16. Kumar, N., Barve, S.A., Chopade, S.S., Kar, R., Chand, N., Dash, S., Tyagi, A.K., Patil, D.S.: Scratch resistance and tribological properties of SiOx incorporated diamond-like carbon films deposited by r.f. plasma assisted chemical vapor deposition. Tribol. Int. **84**, 124–131 (2015). https://doi.org/10.1016/j.triboint.2014.12.001
17. Lanigan, J.L., Wang, C., Morina, A., Neville, A.: Repressing oxidative wear within Si doped DLCs. Tribol. Int. **93**, 651–659 (2016). https://doi.org/10.1016/j.triboint.2014.11.004
18. Zhang, T.F., Wan, Z.X., Ding, J.C., Zhang, S., Wang, Q.M., Kim, K.H.: Microstructure and high-temperature tribological properties of Si-doped hydrogenated diamond-like carbon films. Appl. Surf. Sci. **435**, 963–973 (2018). https://doi.org/10.1016/j.apsusc.2017.11.194
19. Batory, D., Jedrzejczak, A., Kaczorowski, W., Szymanski, W., Kolodziejczyk, L., Clapa, M., Niedzielski, P.: Influence of the process parameters on the characteristics of silicon-incorporated a-C:H:SiOx coatings. Surf. Coat. Technol. **271**, 112–118 (2015). https://doi.org/10.1016/j.surfcoat.2014.12.073
20. Jedrzejczak, A., Kolodziejczyk, L., Szymanski, W., Piwonski, I., Cichomski, M., Kisielewska, A., Dudek, M., Batory, D.: Friction and wear of a-C:H:SiOx coatings in combination with AISI 316L and ZrO_2 counterbodies. Tribol. Int. **112**, 155–162 (2017). https://doi.org/10.1016/j.triboint.2017.03.026

21. Jedrzejczak, A., Szymanski, W., Kolodziejczyk, L., Sobczyk-Guzenda, A., Kaczorowski, W., Grabarczyk, J., Niedzielski, P., Kolodziejczyk, A., Batory, D.: Tribological characteristics of a-C:H:Si and a-C:H:SiOx coatings tested in simulated body fluid and protein environment. Materials **15** (2022). https://doi.org/10.3390/ma15062082
22. Kolodziejczyk, L., Szymanski, W., Batory, D., Jedrzejczak, A.: Nanotribology of silver and silicon doped carbon coatings. Diam. Relat. Mater. **67** (2016). https://doi.org/10.1016/j.diamond.2015.12.010
23. Balestra, R.M., Castro, A.M.G., Evaristo, M., Escudeiro, A., Mutafov, P., Polcar, T., Cavaleiro, A.: Carbon-based coatings doped by copper: tribological and mechanical behavior in olive oil lubrication. Surf. Coat. Technol. **205** (2011). https://doi.org/10.1016/j.surfcoat.2011.01.053
24. Narayan, R.J.: Pulsed laser deposition of functionally gradient diamondlike carbon-metal nanocomposites. In: Diamond and Related Materials, pp. 1319–1330 (2005)
25. Wu, Y., Chen, J., Li, H., Ji, L., Ye, Y., Zhou, H.: Preparation and properties of Ag/DLC nanocomposite films fabricated by unbalanced magnetron sputtering. Appl. Surf. Sci. **284**, 165–170 (2013). https://doi.org/10.1016/j.apsusc.2013.07.074
26. Wang, C., Yu, X., Hua, M.: Microstructure and mechanical properties of Ag-containing diamond-like carbon films in mid-frequency dual-magnetron sputtering. Appl. Surf. Sci. **256**, 1431–1435 (2009). https://doi.org/10.1016/j.apsusc.2009.08.103
27. Manninen, N.K., Ribeiro, F., Escudeiro, A., Polcar, T., Carvalho, S., Cavaleiro, A.: Influence of Ag content on mechanical and tribological behavior of DLC coatings. Surf. Coat. Technol. **232**, 440–446 (2013). https://doi.org/10.1016/j.surfcoat.2013.05.048
28. Batory, D., Reczulska, M.C.-., Kolodziejczyk, L., Szymanski, W.: Gradient titanium and silver based carbon coatings deposited on AISI316L. Appl. Surf. Sci. **275** (2013). https://doi.org/10.1016/j.apsusc.2012.12.088
29. Bociaga, D., Komorowski, P., Batory, D., Szymanski, W., Olejnik, A., Jastrzebski, K., Jakubowski, W.: Silver-doped nanocomposite carbon coatings (Ag-DLC) for biomedical applications - Physiochemical and biological evaluation. Appl. Surf. Sci. **355** (2015). https://doi.org/10.1016/j.apsusc.2015.07.117
30. Bociaga, D., Jakubowski, W., Komorowski, P., Sobczyk-Guzenda, A., Jędrzejczak, A., Batory, D., Olejnik, A.: Surface characterization and biological evaluation of silver-incorporated DLC coatings fabricated by hybrid RF PACVD/MS method. Mater. Sci. Eng. C **63** (2016). https://doi.org/10.1016/j.msec.2016.03.013
31. Moskalewicz, T., Wendler, B., Czyrska-Filemonowicz, A.: Microstructural characterisation of nanocomposite nc-MeC/a-C coatings on oxygen hardened Ti–6Al–4V alloy. Mater Charact **61**, 959–968 (2010). https://doi.org/10.1016/j.matchar.2010.06.005
32. Yetim, A.F., Kovacı, H., Kasapoğlu, A.E., Bozkurt, Y.B., Çelik, A.: Influences of Ti, Al and V metal doping on the structural, mechanical and tribological properties of DLC films. Diam. Relat. Mater. **120**, 108639 (2021). https://doi.org/10.1016/j.diamond.2021.108639
33. Zhou, Y., Li, L., Shao, W., Chen, Z., Wang, S., Xing, X., Yang, Q.: Mechanical and tribological behaviors of Ti-DLC films deposited on 304 stainless steel: Exploration with Ti doping from micro to macro. Diam. Relat. Mater. **107** (2020). https://doi.org/10.1016/j.diamond.2020.107870
34. Zhao, F., Li, H., Ji, L., Wang, Y., Zhou, H., Chen, J.: Ti-DLC films with superior friction performance. Diam. Relat. Mater. **19**, 342–349 (2010). https://doi.org/10.1016/j.diamond.2010.01.008
35. Vitu, T., Escudeiro, A., Polcar, T., Cavaleiro, A.: Sliding properties of Zr-DLC coatings: the effect of tribolayer formation. Surf. Coat. Technol. **258**, 734–745 (2014). https://doi.org/10.1016/j.surfcoat.2014.08.003
36. Escudeiro, A., Polcar, T., Cavaleiro, A.: A-C(:H) and a-C(:H)-Zr coatings deposited on biomedical Ti-based substrates: tribological properties. In: Thin Solid Films, pp. 89–96 (2013)
37. Banerji, A., Bhowmick, S., Alpas, A.T.: High temperature tribological behavior of W containing diamond-like carbon (DLC) coating against titanium alloys. Surf. Coat. Technol. **241**, 93–104 (2014). https://doi.org/10.1016/j.surfcoat.2013.10.075

38. Dai, W., Gao, X., Liu, J., Wang, Q.: Microstructure, mechanical property and thermal stability of diamond-like carbon coatings with Al, Cr and Si multi-doping. Diam. Relat. Mater. **70**, 98–104 (2016). https://doi.org/10.1016/j.diamond.2016.10.017
39. Cao, H., Bai, M., Inkson, B.J., Zhong, X., De Hosson, J.T.M., Pei, Y., Xiao, P.: Self-healing WS_2 tribofilms: an in-situ appraisal of mechanisms. Scr. Mater. **204** (2021). https://doi.org/10.1016/j.scriptamat.2021.114124

Chapter 6
Methods of Testing Friction and Wear of Coatings

Abstract This chapter is devoted to a discussion of tribological research. The chapter presents a systematics of experimental research and the structure of dependencies between input and output parameters for a typical tribological situation under investigation. It also presents the most common methods for testing the friction and wear properties of carbon coatings. Furthermore, it discusses equipment diagrams, measurement methodology, and experimental data processing and presentation.

The long-term and trouble-free operation of friction contacts in a wide range of operating conditions is contingent upon the correct selection and matching of mating materials, the appropriate design of the friction node, and adherence to operating specifications. It is therefore evident that an understanding of the tribological parameters of the materials used in designed or optimised friction pairs represents a fundamental and, in many cases, a critical element. The consequences of a failure in a friction pair can be significant potentially leading to the failure of the individual components, as well as the entire assembly of machines and devices. Such failures can pose an immediate threat to health and life, and in the long term, may affect the continuity of production, deliveries, or other services provided. In the aerospace industry, it is standard practice to employ only proven and reliable solutions, the effectiveness of which has been confirmed by the results of laboratory tests conducted over many years, as well as industrial research. Figure 6.1a, b illustrate an example of crankshaft pans wear in an internal combustion engine, while Fig. 6.2 depicts a seized pair of crankshaft pans.

A deficiency in lubrication and the resultant deterioration of sliding surfaces have the potential to cause the piston-crank system to become irreversibly immobilised, with the possibility of inflicting permanent damage to the engine structure (Fig. 6.3).

The acquisition of a comprehensive dataset on the frictional and anti-wear characteristics of interacting materials is only possible through empirical investigation. The fundamental objective of these experiments is to elucidate the phenomena occurring in a friction contact. Their accurate identification, accompanied by a qualitative and quantitative description, enables the determination of the nature and progression of friction processes, as well as an analysis of their effects, namely resistance to motion and progressive wear.

110   6 Methods of Testing Friction and Wear of Coatings

Fig. 6.1 An example of damage to the crankshaft pans

Fig. 6.2 The pair of pans seized on the crankshaft journal

Fig. 6.3 An example of the damage caused to an internal combustion engine block by a failure of the piston-crank system

6.1 Tribological Investigations

Experimental tribological research in its broadest sense, can be categorised into two distinct groups:

- in-service testing—conducted under real conditions on actual facilities. The primary advantage of this approach is that the results obtained can be directly applied in practical settings. However, it should be noted that the technique is associated with several disadvantages, including the high cost, the extended time requirement for testing, and the generally significant measurement difficulties.
- laboratory tests—on models of real objects or on isolated hardware units. However, these tests have similar drawbacks to in-service testing. Another type of laboratory testing is that conducted using standardised material samples (friction pairs). Their use is not associated with high testing costs, apart from, of course, the cost of purchasing the apparatus itself. They are characterised by ease of measurement and the highest repeatability of results. Furthermore, these tests permit the comparison of results obtained in other test centres, on the condition that all essential test parameters have been defined.

A customary exemplar of a tribological research situation is presented in Fig. 6.4. It comprises the tribological contact under scrutiny (i.e. the sample, counterbody, lubricant, and environment), specific controlled input parameters, and controlled output quantities.

The input operating quantities include:

- Motion parameters (type of motion, load, speed, vibration, sliding distance, temperature, time).

The friction contact characteristics include:

- Sample under test (substrate material, substrate material processing history and mechanical properties, coating manufacturing method, coating chemical composition, geometrical surface structure parameters and mechanical properties)
- Counterbody (material, dimensions, geometrical surface structure parameters, mechanical properties)
- Test environment (type and chemical composition of working atmosphere, relative humidity, temperature, type of lubrication, lubricant type and flow rate).

The output quantities can be classified into three categories:

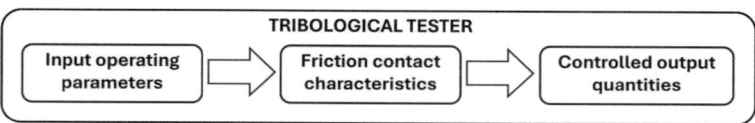

Fig. 6.4 Schematic of a typical tribological research situation

1. Directly measured quantities, which include linear wear, friction force/moment, temperature near the friction contact, number of revolutions/cycles, and vibration,
2. Directly determined quantities, which include wear intensity, coefficient of friction, and sliding distance,
3. Additional quantities determined during or after the test using complementary test techniques. These are discussed in more detail later in this chapter.

In addition to the basic input and output quantities mentioned above, a number of additional uncontrolled factors occur in tribological systems, the most important of which are [1]:

- The appearance of wear products in the contact zone
- Change in chemical properties
- Change in physical properties
- Change in the mechanical properties of the friction pairing materials
- Change in the microgeometry of the surface layer
- Change in stress distribution in the surface layer.

The schematic dependence of the input and output quantities in the tribological tester is shown in Fig. 6.5.

A significant factor that is not always given due consideration is the relationship between output and input quantities. To illustrate, an increase in wear can lead to alterations or the generation of vibrations, thereby modifying the dynamics of the load and acting as a forcing mechanism. A similar effect can be observed in the case of heat generation due to frictional phenomena. This may result in local phase transitions and microstructural changes, which affect the elastic properties. These, in turn, determine the deformations and surface pressures, the physico-chemical changes that determine the type of interaction between the surface layers of the mating parts, or finally, the changes in lubricant viscosity [1].

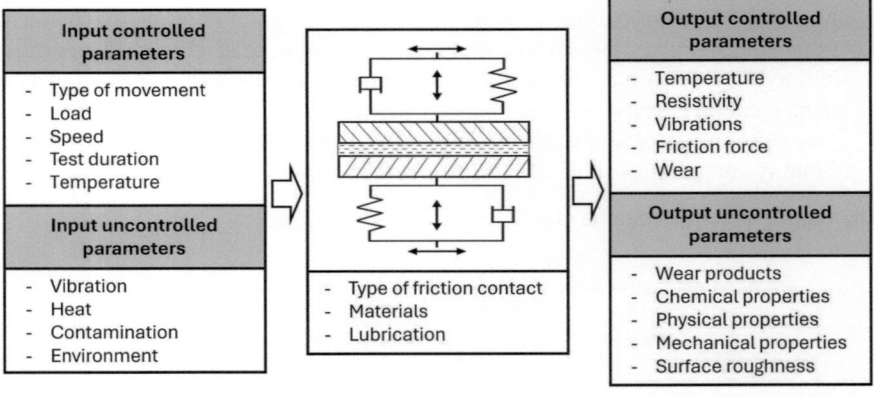

Fig. 6.5 Dependency structure of input and output quantities in the tribological tester

It can thus be concluded that wear resistance and resistance to motion are not merely characteristics of individual materials or even properties of the friction pair; rather, they are properties of the entire tribological system, and depend on the operating conditions under which it is used. It is therefore evident that a more profound comprehension of the tribological characteristics of a specific friction pair can only be attained through the implementation of diverse methodologies and techniques, encompassing an array of test parameters. Consequently, for comprehensive in depth analysis of tribological properties of friction contacts advanced apparatus solutions are employed, offering high-precision measurements and the possibility of comparing the results with those of other research centres. Nevertheless, in certain instances, the undertaking of meaningful comparisons is rendered arduous, if not wholly unfeasible, by an absence of pertinent information with regard to the experimental conditions, the processes of sample preparation, and other pertinent factors [2].

6.2 Tribotesters

The principal means of acquiring knowledge of the tribological properties of materials is through the use of laboratory devices, specifically tribotesters. The configuration of the friction association and the test conditions (normal force, sliding speed, pressure, temperature, atmosphere, lubricant) influence the design of the tribotester and the number and range of parameters recorded, which are selected directly for the specific test. In practice, however, there are no dedicated methods for testing the tribological properties of coating materials. These solutions are essentially adaptations of standardised methods employed in tribological testing of solid state materials, for which normative parameter ranges and test input conditions have been developed. The test results obtained using these methods are a reliable source of information that can be successfully used in the design or optimisation processes of DLC-based friction contacts. However, it is noteworthy that several publications in the extant literature describe the tribological properties of carbon-based coatings tested using individually selected input quantities (applied load, sliding speed, etc.) that deviate from those defined by relevant standards. The most common model friction assemblies employed in testing, classified according to the nature of contact between mating elements, are illustrated in Fig. 6.6. These configurations encompass the following: (i) point contact (ball-on-disc); (ii) linear contact (block-on-ring); (iii) surface contact (pin-on-disc).

The aforementioned types of contact predominantly manifest in initial conditions, as testing may result in a transition from point or linear contact to surface contact due to elastic or plastic deformation, in addition to progressive wear of the sliding pair components.

The following examples illustrate the test equipment most commonly employed to study the tribological properties of coating materials, including low-friction carbon-based materials.

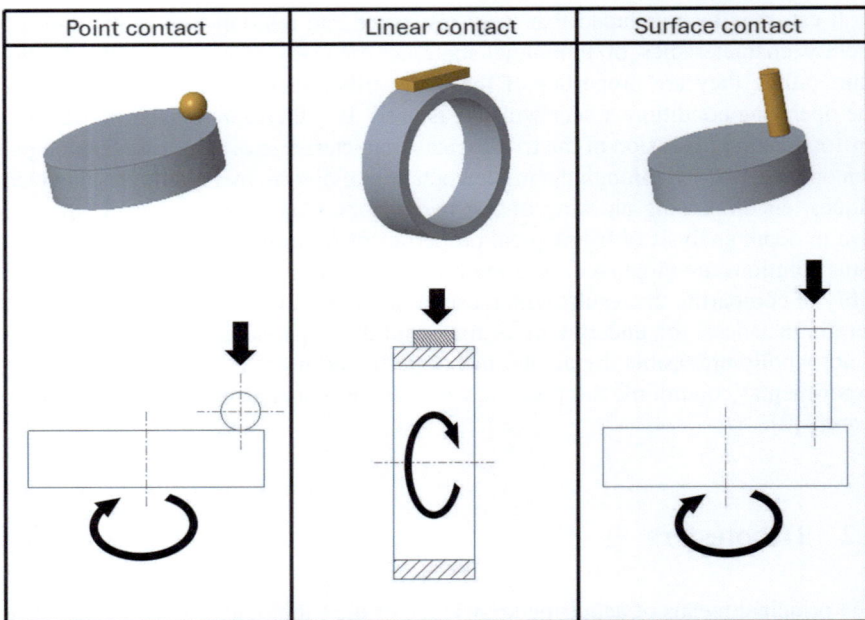

Fig. 6.6 Schematic representation of friction contacts taking into account the nature of contact between mating surfaces

6.2.1 A Device with a Pin/Ball-on-Disc Configuration

The most common apparatus employed for the assessment of tribological characteristics of low-friction coatings is a testing apparatus comprising a pin-on-disk or ball-on-disk configuration. This device is utilised for the evaluation of material wear, the assessment of coating durability and sliding properties, and the analysis of the influence of load, sliding speed, the presence of lubricant and controlled operational atmospheres, including elevated or low temperatures, on machine components subjected to significant stress [1, 3–8]. Furthermore, these devices are equipped with temperature measurement systems in the vicinity of the friction node, as well as a displacement sensor to determine the total linear wear of the friction pair elements. The relative movement of the friction contact can be accomplished by rotation of the spindle on which the tested sample is mounted (see Fig. 6.7) or by reciprocating movement through a movable table to which one of the friction pair elements is attached.

A comprehensive review of the extant literature and the author's extensive experience suggests that this is the most prevalent type of tribological tester, constituting fundamental instrumentation in scientific laboratories worldwide. The popularity of this type of tribological tester can be attributed to several factors, including the straightforward method of preparing samples and the extensive availability of

6.2 Tribotesters

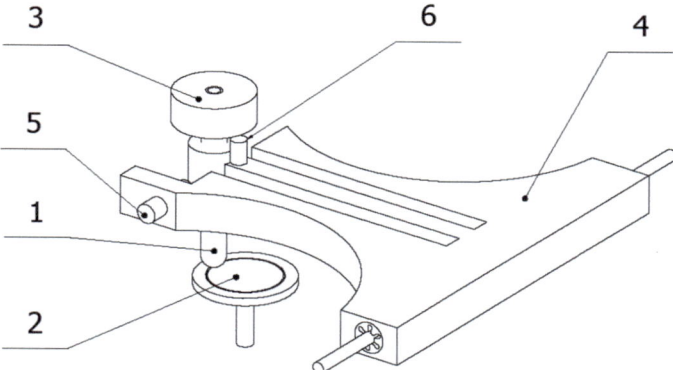

Fig. 6.7 Schematic diagram of the device with pin-on-disc configuration: (1) counterbody holder with attached pin or ball, (2) rotating spindle with a tested sample, (3) replaceable weight constituting load on the friction contact, (4) tribotester arm with attached counterbody holder, (5) friction force sensor, (6) displacement sensor

compatible components, such as pins or balls. However, its primary advantage lies in the high precision of the geometrical characteristics of the sliding pair, which is a consequence of the absence of mutually matching dimensions. This provides a high degree of confidence in the value of the applied unit pressure and its uniform distribution over the sample surface. The simplicity of the design and principle of operation also allows for modifications that do not require specialist knowledge, such as adapting the system for testing in liquids or controlled temperatures (Fig. 6.8) [9–11].

In addition to macro-scale testing, there are numerous microtribotometer solutions that operate on the same principle, albeit at significantly lower loads, ranging

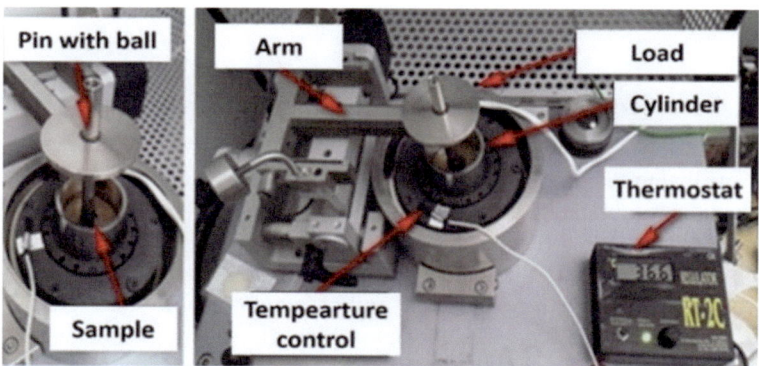

Fig. 6.8 The ball-on-disc test rig modified to allow for the testing of friction contacts in a simulated body fluid environment, specifically through the modification of the specimen holder. Reprinted with permission from [9]

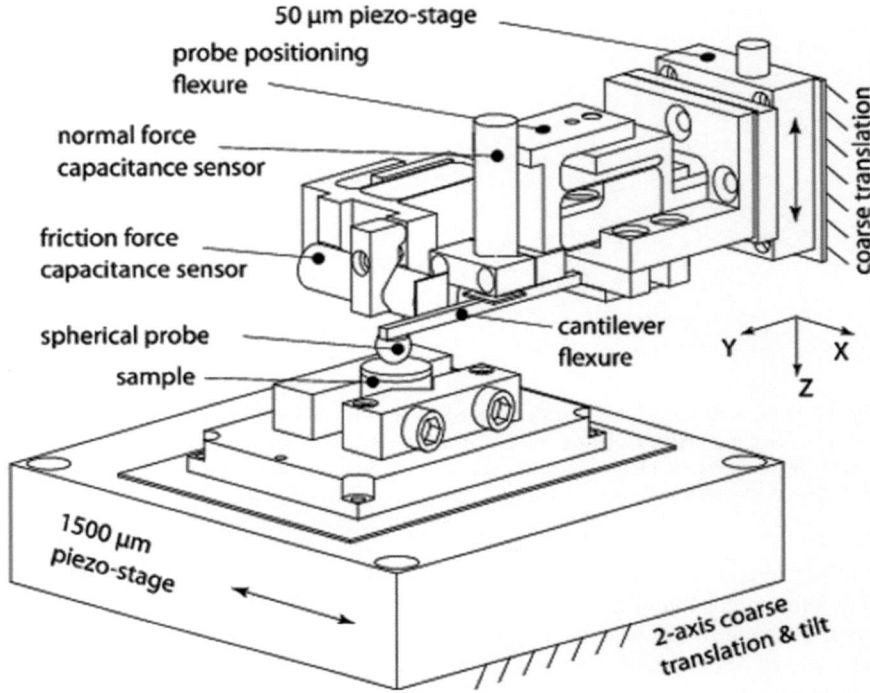

Fig. 6.9 Schematic of a microtribometer. Reprinted with permission from [14]

from single to tens of mN. An illustrative example of a microtribometer design is presented in Fig. 6.9. These methods facilitate a more profound comprehension of the tribological mechanisms occurring at the submicrometric scale of contact [12, 13].

6.2.2 A Device with Block-on-Ring Configuration

Devices ustilising the block-on-ring configuration are predominantly employed for the evaluation of lubrication characteristics, encompassing the assessment of plastic greases, oils and solid lubricants, in addition to the determination of wear resistance in metals and polymers. In the evaluation of low-friction coatings, these devices are employed to ascertain their resilience to galling, predominantly for highly stressed machine component surfaces. The apparatus allows for the implementation of tests in both dry and lubricated sliding contact, with progressive or oscillatory motion at speeds and amplitudes across a wide range, as well as loads. Tests may be conducted in both concentrated linear and distributed contact. However, as with other methods of tribological testing of friction pairs, it is evident that the nature of the contact will evolve throughout the course of the test, contingent on the progressive wear of the friction node components. The schematic representation of the block-on-ring contact apparatus is illustrated in Fig. 6.10. The axis of rotation is situated above the lubricating fluid mirror, at a distance less than the radius of the roller. This configuration enables the rotation to draw the lubricant into the contact area [1, 15].

In the study of low-friction carbon-based materials, the method has also been employed for the analysis of the wear resistance of coatings in the presence of a third body, in water–oil emulsion with addition of different size abrasive silica [16], as well as water with alumina particles [17].

In general terms, the determination of the wear and friction characteristics of DLC coatings or their modified variants using the ball-on-disc and block-on-ring methods can be reduced to a single, fundamental process: the performance of a friction cycle at a given load on a defined friction pair, with the subsequent measurement of the friction force (coefficient of friction) and temperature, and in some cases linear wear. Subsequent testing may be conducted to examine the geometry, morphology, composition, and chemical structure of the wear track and wear scar on both components of the friction pair, as well as the mass loss value. The aforementioned methods thus permit the quantification of the tribological characteristics of the tested friction pairs, facilitating the clear definition of universally accepted and comparable functional parameters, such as the coefficient of friction or the wear rate. The latter encompasses the volumetric or mass loss of the counterbody material in relation to the load and the sliding distance. However, it should be noted that these properties are largely characteristic of a specific tribological system.

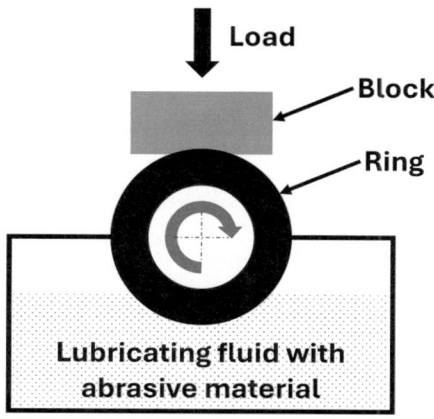

Fig. 6.10 Schematic of the device with block-on-ring configuration

6.2.3 Nanoscale Tribology

In the event that our concerns pertain to atomic and molecular interactions at the nanoscale, occurring upon the frictional contact of materials, what methodology should be employed? The physical origins of tribology require the examination of well-defined interfaces. Consequently, experiments often focus on single asperity contacts (i.e. the contact area remains constant during the whole test). The prevailing opinion in the AFM literature suggests that, in many solid–solid nanocontacts below the wear threshold, friction is proportional to the true contact area (i.e. the number of interfacial atoms) [18, 19]. A nanoscale single asperity contact is formed between the AFM tip and the sample's surface, and normal and lateral forces and displacements are measured with atomic-scale resolution. A schematic of a typical AFM instrument is presented in Fig. 6.11. Friction and wear analysis at the macroscale typically pertains to conditions where the contact states are complex (i.e., multi-asperities contact or three-body contact) and the wear is severe. Consequently, the actual contributions of phase transition and surface property evolution in the tribological behaviour of DLC coatings may be obscured by the simultaneous occurrence of multiple interactions [20–22].

This enables an in-depth analysis of the friction of carbon-based materials at the atomic or molecular level, thus allowing for the avoidance of significant complexities brought about by plastic deformation or wear [23]. Conversely, it opens up a range

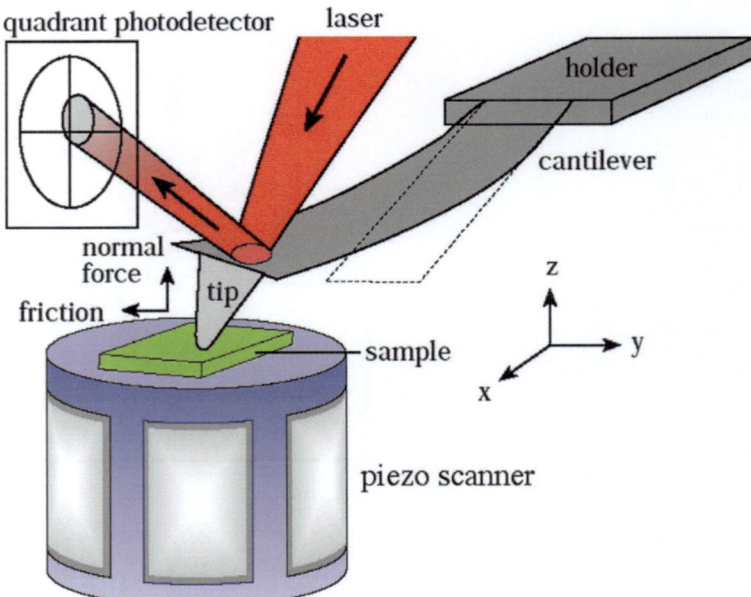

Fig. 6.11 Schematic of a typical AFM instrument. Reprinted with permission from [18]

of possibilities for the analysis of wear and friction. This includes the analysis of extremely thin diamond-like carbon films, typically applied to surfaces involved in the magnetic recording head-disk interface [24, 25] or microelectromechanical systems (MEMS) [26, 27]. Equally important are studies on the tribological properties of carbon coatings intended for biomedical applications, in particular those doped with other elements that, under the influence of wear processes or phase changes, may have a negative impact on the human body [28]. Commercially available AFM cantilevers are usually equipped with diamond, Si or Si_3N_4 tips with only a few other tip coatings available (including DLC).

6.2.4 Fatigue Analysis by Micro and Nanoimpact Testing

Diamond-like carbon coatings have emerged as a promising material for a wide range of industrial applications. In addition to resisting friction and wear caused by abrasion or adhesion, these coatings are capable of withstanding erosive wear or dynamic loading during service. However, they are susceptible to failure due to fatigue processes. In order to ensure the reliability of these coatings, it is essential to conduct laboratory tests that accurately simulate future working conditions. In this particular case, the focus should be on assessing toughness, adhesion, and impact resistance. Impact testing (on both micro and nano scales) has emerged as a technique for characterising fracture resistance of hard coatings. The data on fatigue performance can be used in coating development processes, including optimisation of the chemical composition and deposition parameters for coating, improved adhesion and durability [29]. The schematic view of the pendulum-based nanoindentation device is presented in Fig. 6.12.

The test procedure generally involves the application of the spherical or conical indenter, which is driven by the solenoid or other mechanical system to produce repetitive impacts on the coating surface. The applied load and frequency may vary depending on the test setup and assumed methodology. The test conditions and duration up to the film damage initiation, are considered in order to determine the critical stresses associated with the coating fatigue strength [31]. Furthermore, the influence of maximum normal impact load, absorbed energy, and contact impulse on the residual impact crater volume/depth of DLC coating is analysed [32]. In this regard, McMaster et al. [33] proposed the following features of the impact depth maps in the coatings that can be used to compare their performance across the selected loads:

- quasi-static depth.
- the depth of the first true impact.
- the depth of the final impact.
- the ratio of final depth to initial depth normalised by the initial impact depth. This parameter shows the relative level of fatigue (depth increase due to crack formation) between each loading step.

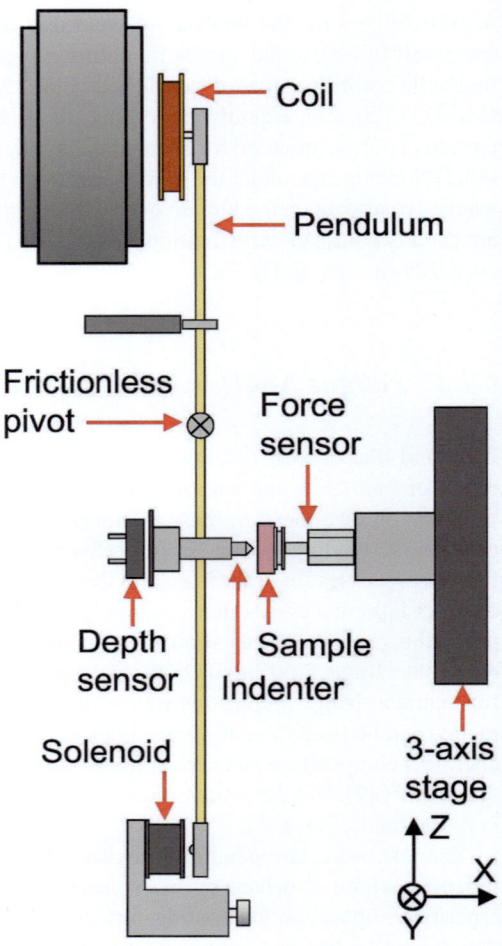

Fig. 6.12 Schematics of pendulum-based nanoindentation device. Reprinted with permission from [30]

The impact testing method has been demonstrated to be a valuable tool in the analysis of hard coatings, particularly in the context of tool modification [34]. Its applications extend to the assessment of fatigue resistance under repetitive loading, such as the impacts of erosive particles. In this regard, a significant correlation has been observed between the relative depth increase in instrumented impact and substrate exposure in erosion [33]. This underscores the necessity for a multi-faceted research approach, encompassing diverse and complementary techniques, to thoroughly characterise the mechanical properties of coatings and evaluate their performance.

6.2.5 Microabrasion Wear Testing (Calotest)

The ball-cratering method is a widely employed technique for the evaluation of anti-wear and low-friction coating materials in terms of their resistance to abrasive wear. The method entails the removal of coating material and substrate through the utilisation of a rotating steel ball of known radius and mass, along with a specified rotational speed. During the test, a fine-grained abrasive slurry, comprising SiC, Al_2O_3, diamond, or other materials, is typically introduced into the friction node, contingent on the objectives of the test. The schematic of the apparatus employed for measuring abrasive wear is illustrated in Fig. 6.13.

As a consequence of the rotation of the sphere, a continuous layer of the abrasive suspension is formed on its surface, with the weight and rotation of the ball transferring normal force and tangential force to the abrasive particles. These particles interact with the surface under examination, causing abrasive wear. The result of the test is a circle-shaped pattern on the specimen, which can be quantified using an optical microscope. Based on the measurement of the crater diameter, the volume of the coating can be calculated. However, it is more appropriate to use Archard's wear equation for homogeneous materials to determine the coefficient of wear, which is the volume of wear normalised with respect to the test parameters (load and friction distance) [35, 36]:

$$k = \frac{V_w}{sF} \quad (6.1)$$

where:

k wear coefficient
V_w volume of the removed coating material

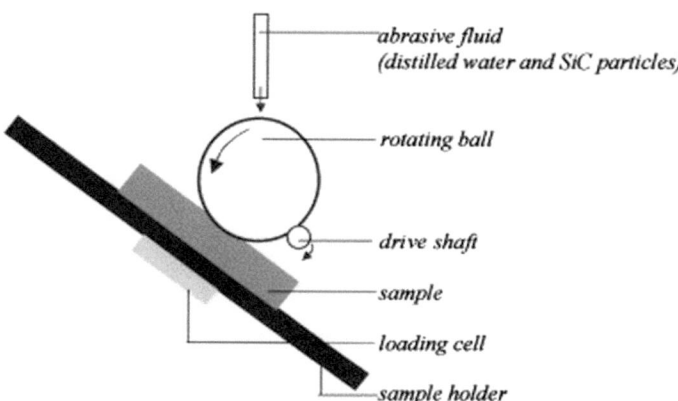

Fig. 6.13 Schematic of the apparatus used for measuring the abrasive wear. Reprinted with permission from [35]

s   sliding distance
F   normal force (resulting from the weight of the ball).

A schematic representation of the boundary conditions for the measurement of the volume of removed coating material is shown in Fig. 6.14.

The volume of the crater can be determined from the following relationships [37]:

$$V_w = \frac{1}{3}\pi b^2 (3r - b) \tag{6.2}$$

$$b = r - \sqrt{r^2 - \frac{d^2}{4}} \tag{6.3}$$

where:

r   ball radius
b   depth of the callote
d   callote diameter.

It is imperative that the depth of the callote during testing is less than the thickness of the coating being tested, in order to prevent the coating from being completely abraded. In the event that this condition is met, the abrasion mark on the surface of the test sample will manifest as concentric circles. Consequently, the procedure for determining the wear factor will assume a slightly more intricate but viable form [38]. However, this scenario offers a further opportunity to measure the thickness of the coating undergoing examination. Figure 6.15 illustrates an example of an abrasion trace displaying a distinct coating perforation and discernible substrate material.

The thickness of the coating (t) can be determined based on simple equation:

$$t = \frac{(D^2 - d^2)}{8r} \tag{6.4}$$

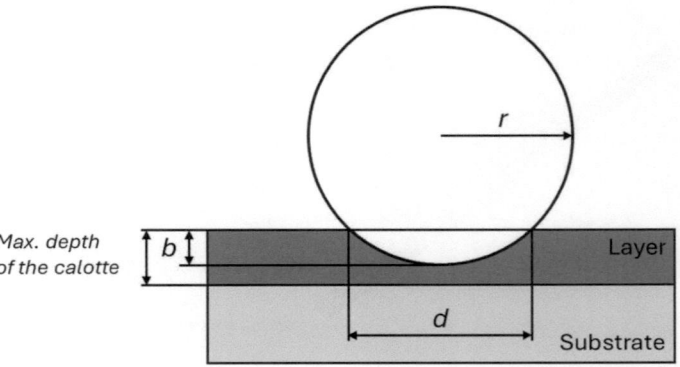

Fig. 6.14 Boundary condition for wear measurements

Fig. 6.15 Coating thickness measurement on a flat sample

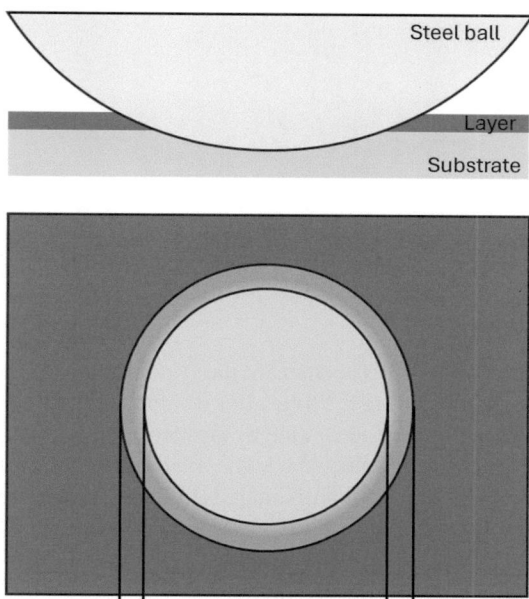

where:

D   outer calotte diameter
d   inner callote diameter
r   ball radius.

6.2.6 A Device with Pin and Vee-Block Configuration

The pin and vee block test device is utilised for the evaluation of materials' wear resistance and sliding properties at friction nodes, encompassing plastic lubricants and oils. The device's primary advantages include its simplicity of construction, the uncomplicated shape of the specimens, and the low requirements for accuracy. In the context of testing low-friction coatings, this method can be employed to ascertain their wear characteristics and resistance to galling. Tests can be carried out both under dry friction conditions and in the presence of a lubricating fluid. The schematic of the pin and vee block configuration is presented in Fig. 6.16.

The test method involves subjecting a rotating pin to a specified force through two prismatic blocks. The entire contact surface may be immersed in a temperature-controlled lubricating fluid. To ascertain the wear characteristics, an indirect measure of the total wear of the specimens and the counterbody can be obtained by determining

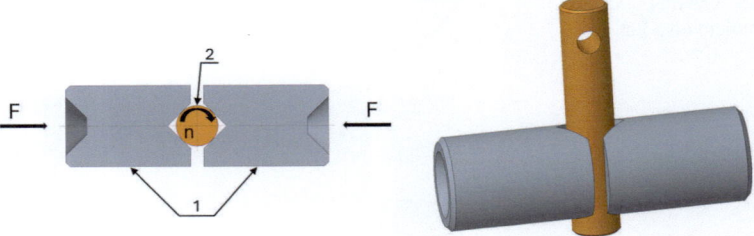

Fig. 6.16 Schematic view of pin and vee block tribosystem: (1) vee block, (2) pin

the value of displacement of the system that applies the compressive force to the pads, or by measuring the time after which the total wear of the test coatings has occurred. During the test, the value of the force applied to the pads is increased in increments of a constant value (the same at specified intervals). In contrast, in the case of the galling test, the load is increased linearly, and the measure of galling resistance is the value of the load on the friction node at which the node seizes. By continuously measuring the friction torque, the process of identifying the moment of total wear or seizure of a friction node is relatively simple [1, 39].

6.2.7 A Device with Three Rolls and Cone Configuration

The apparatus, comprising three rollers and a cone, has been constructed for the purpose of evaluating the wear and seizure resistance of materials operating within friction nodes. A schematic representation of the analysed tribosystem is provided in Fig. 6.17. The test procedure is based on the measurement of friction under specified conditions (rotational speed, contact load, lubrication method) between three stationary roller specimens and a rotating cone-shaped counterbody. The linear wear of the friction node or the load at which seizure occurred is determined. The test is carried out in concentrated point contact, at constant speed and load (with the possibility of varying its value between tests). During the test, the resistance to rotary motion and the linear displacement of the friction node loading system are recorded [1, 40]. This methodological framework facilitates the execution of comparative studies and enables the determination of the tribological properties of the tested coatings under conditions that closely resemble real working environments.

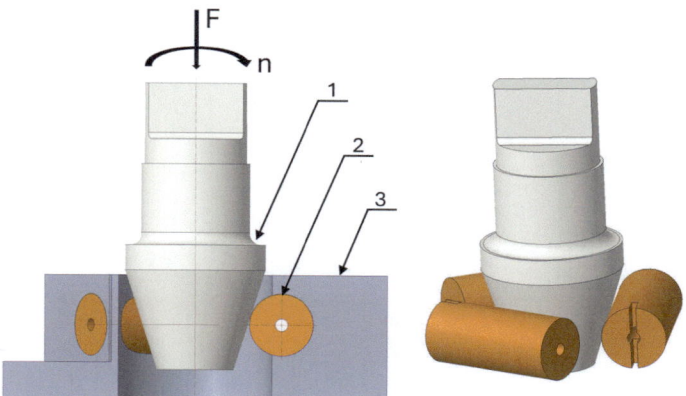

Fig. 6.17 Schematic of three rolls and cone tribosystem: (1) cone; (2) roll; (3) holder

6.2.8 Three-Ball-on-Rod Rolling Contact Fatigue (RCF) Testing Machine

The roller-type RCF testing device is a widely utilised apparatus employed for the evaluation of the rolling contact fatigue performance of technical coatings. These devices facilitate the simulation of the rolling/sliding contact configurations, for instance in the context of gear and cam/tappet applications. The balls are positioned at a separation of 120° within a metal retainer, facing a rotating test rod. The application of load is facilitated by pre-stressed springs acting upon two tapered cups. The test is conducted under constant lubrication conditions, with the vibration of the entire system being constantly measured by sensors. Should failure occur (due to spalling, delamination or pit formation) this is defined by the increase in the vibration amplitude and is registered by the computer. The time for a fatigue failure or fatigue spalling to develop on the rod is recorded, and the number of stress cycles is calculated. Finally, Weibull distribution is used to analyse the fatigue data [41, 42]. The view and schematic diagram of the contact rolling fatigue tester is presented in Fig. 6.18.

Despite the overall very good resistance of DLC coatings to rolling contact fatigue (RCF), the mechanism of their failure remains to be fully elucidated. Due to the minimal thickness of the carbon coatings that were examined (<1 µm), it is improbable that the initial spallation would lead to an increase in vibration amplitude, which would typically serve as an indicator of RCF failure. Consequently, the test is conducted at a fixed interval. Systematic RCF tests, in conjunction with microscopic examination following various test intervals, have demonstrated that micro-polishing of DLC coating may significantly contribute to prolonging fatigue life. Spalled coating debris may be present within the contact region and initiate polishing of the surfaces due to its mixing with the test lubricant. It is hypothesised that this polishing effect reduces asperity loading and hinders the initiation of surface

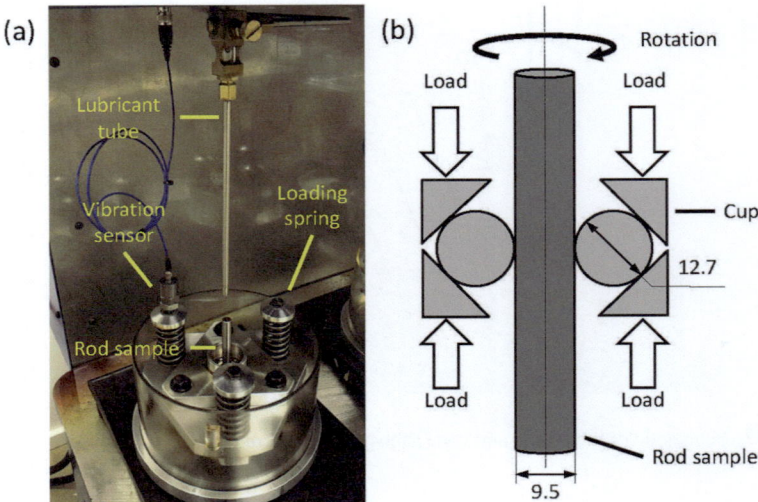

Fig. 6.18 a View of the three-ball-on-rod RCF tester, **b** Schematic diagram of the rolling contact fatigue tester. Reprinted with permission under Creative Commons CC-BY from [43]

crack formation. In addition, an order-of-magnitude increase in the fatigue life was observed. Furthermore, there was no evidence of coating delamination, suggesting that failure occurred as a result of coating wear followed by conventional spalling [41, 42, 44].

The aforementioned considerations pertain primarily to standardised tribological testing methodologies, which, to a certain extent (although this is intentional), represent a significant reduction in the complexity of tribological systems. However, this simplification does not fully align with the actual tribological interactions observed in contemporary machine and device designs and kinematic solutions. The majority of the test procedures employed are designed to facilitate a comprehensive evaluation of the performance of the tested friction contact within a relatively short timeframe and without significant financial expenditure, often achieved through comparative assessments. Despite their simplicity, low level of complexity and ease of sample preparation, these tests provide an important source of information for further optimisation work or laboratory and in-service testing, with the ultimate goal of a final industrial application confirmed by reliable test results. Given the extensive existing literature on the use of these basic research methods to analyse the tribological properties of carbon coatings, the author has limited himself to a general presentation of them without going into unnecessary detail, let alone a review of the values of friction coefficient, wear, galling or rolling contact fatigue obtained for carbon-based coatings. It is also important to note that another group of tribological tests involves the use of dedicated devices (not necessarily standardised) that are designed to represent real tribosystems operating under rolling and/or sliding friction. This topic will be explored in greater detail in the following subsections.

6.2.9 Micropitting Studies of Coated Surfaces in Rolling/Sliding Contacts

The ongoing demand for reduced weight and enhanced performance in machines and devices employed in the agricultural, offshore wind and tidal power generation, aviation, space technology and automotive industries (just to name few) necessitates gearboxes capable of withstanding high power densities while ensuring prolonged operational lifespan. These rigorous operating conditions give rise to substantial local stresses within the cooperating surfaces of gear pairs and the rolling elements of bearings within the gearbox. The elevated contact pressures result in accelerated surface damage accumulation, leading to the nucleation and propagation of micro-cracks, and consequently, micropitting [45–47]. In conditions of cyclic load and lubrication, material fatigue is characterised by the formation of surface cracks. These cracks then propagate only a few micrometres from the surface into the subsurface region. They rarely reach significantly deeper than 10 μm, which is approximately equivalent to the height of the asperities of the surface geometrical structure. Cracks that have already initiated growth may merge, forming microscopic pits. The micropits may appear after a relatively short period of operation (in some cases less than a million cycles). The failure is manifested by increased noise and vibrations caused by the deviation of the nominal dimensions of the element as a result of micropitting. The cracks propagate into the depth of the steel at a shallow angle, usually less than 30°, and they can merge, resulting in a continuous loss of material, thus the profile deviation [48, 49]. A schematic representation of the micropitting process and its further evolution is shown in Fig. 6.19.

A multitude of factors have the capacity to exert influence upon the micropitting process, including but not limited to the type of lubricant utilised, the presence of contaminants, the temperature at which the process is conducted, contact stresses, the hardness of the materials involved, the sliding speed, and the surface roughness. In the event of inappropriate conditions, the profile deviation gains due to micropitting and sharp-edged transitions are formed. This phenomenon is known to result in the formation of stress concentrations, which, in turn, give rise to the development of

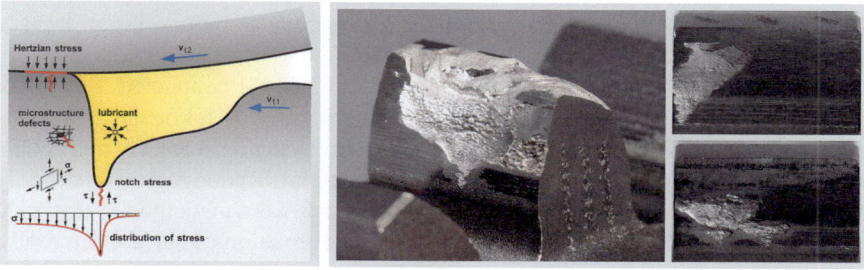

Fig. 6.19 Micropitting area transition and test gears with pitting. Reprinted with permission from [49]

additional cracks. The presence of such signs on the tooth surface has been shown to increase both wear and the probability of pitting occurrence [49].

The reliability of gears and bearings working under sliding-rolling friction conditions can be noticeably improved by maintaining the formation and evolution of microcracks and wear at an acceptable level. This can be achieved by depositing technical coatings on the surface. The high hardness of these coatings (in comparison to the bulk gear material) and low friction coefficient result in minimised interaction between surface asperities in the contact region. While the change in coating thickness does not have a major impact on the contact load ratio, oil film thickness, and coefficient of friction, it has a significant effect on the maximum surface tensile stress and interfacial shear stress, especially for high-friction and low-speed conditions. Consequently, the implementation of thin, low-friction coatings has been demonstrated to enhance the tribological performance of bearings, thereby extending their service life, particularly in harsh working conditions characterised by inadequate lubrication and severe asperity contact [50].

One method of testing the micropitting of carbon coatings on the surface of rolling elements of bearings is to use a test rig with three counter rings and a roller tribosystem in a planetary configuration (see Fig. 6.20). The counter rings and the central roller specimen are driven by different shafts, enabling tests to be conducted under varied slide-to-roll ratios, i.e. the ratio between the sliding speed and the entrainment speed, according to the following formula:

$$SRR = \frac{U_{ring} - U_{roller}}{0.5(U_{ring} + U_{roller})} \quad (6.5)$$

where:

SSR   slide to roll ratio
U_{ring} rotational speed of the counter ring
U_{roller} rotational speed of the roller.

The load is applied to the upper counter ring, whereas the lower counter rings are immersed in the lubricant sump. The rotational speed of the former circulates lubricant around the central roller. The test rig is equipped with a torquemeter and an accelerometer, both of which are utilised for continuous monitoring of surface damage in the central specimen along the test duration.

Using the presented test rig, Zapata Tamayo et al. [47] conducted a series of micropitting tests on DLC layers in the presence of various lubricants. The tests were conducted under conditions of contact pressure ranging from 1.9 to 2.5 GPa and a specific shear stress (SSR) of 5%, within a mixed lubrication regime where only the roller component was coated. These test conditions were meticulously designed to promote the nucleation of micro-cracks and the formation of surface micropits. The authors of the study posit that in coated systems, rolling contact fatigue leads to the generation of localised micro-spallation of the coating in the contact area, as opposed to the formation of wide V-shaped microcracks that result in deep pits.

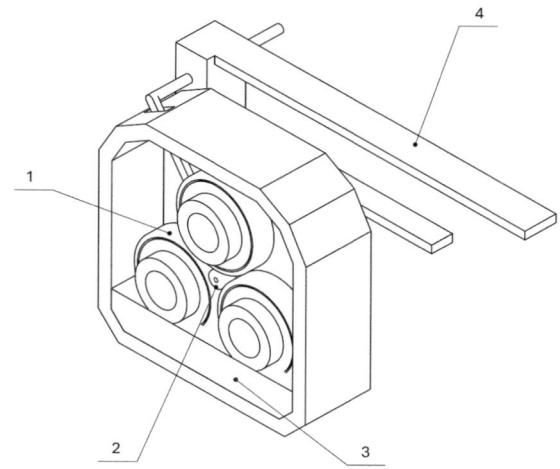

Fig. 6.20 Schematic representation of a micro pitting test rig [47]; (1) counter-ring; (2) roller (specimen); (3) sump-volume; (4) loading arm

The test results indicate that amorphous carbon coatings enhance the contact fatigue resistance for lubricants that induce mild-wear as a dominant failure mode. Consequently, this form of wear leads to the removal of superficial pits, which in turn give rise to deep grooves resulting from the interaction between surface asperities. The progressive nature of abrasive wear processes serves to inhibit the coalescence of multiple micropits (through the propagation of micro-cracks) and their subsequent evolution into macropitting. In comparison with uncoated surfaces, DLC coatings have been demonstrated to be effective in preventing the initiation and propagation of micro-crack nucleation and coalescence of micropits. This results in enhanced fatigue life and a substantial reduction in volume loss during rolling-siding contact. However, in the case of lubricants that induce micropitting as the dominant fatigue mode, DLC coatings are prone to failure due to micro-spallation and subsequent exposure of the steel surface to the formation of macropits.

Similar rest results obtained for 1.79 GPa contact pressure and SSR of 40% were reported by Singh et al. [51]. The authors of the study suggest that coating merely one element of the material pair is sufficient to delay the onset of surface fatigue in test rollers by more than 100 million cycles.

In a study by Michalczewski et al. [52], a device for analysing micropitting of DLC coatings was presented. This device is less complicated than others currently available, yet it is still effective. The T-02U universal four-ball testing machine was adapted by the authors, who substituted the top ball with a cone modified with W-DLC/CrN coating (see Fig. 6.21). The tests were performed under constant load and sliding speed, while the contact zone between the balls and cone was immersed in oil. The experiment is continued until pitting is detected, with a total of 24 tests being conducted. Subsequently, the Weibull distribution is utilised to determine the L_{10} parameter, which represents the time at which 10% of the test cones are expected to fail, based on the time to failure values recorded after all 24 runs.

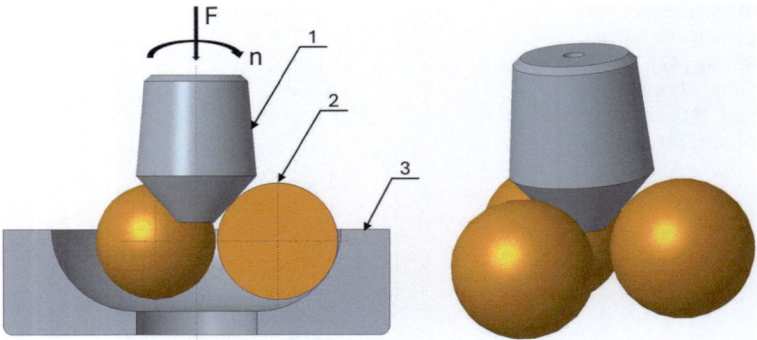

Fig. 6.21 A schematic of three balls-cone tribosystem: (1) cone, (2) ball, (3) race

The findings reveal that, in contrast to earlier reports, the deposition of carbon coatings substantially reduces the resistance to pitting, irrespective of the type of oil employed. The authors hypothesise that this phenomenon is attributable to phase transformations in the substrate material induced by the temperature during the carbon-based layer deposition process, a hypothesis that was corroborated by the decline in microhardness of the steel substrate following the coating synthesis process. However, in the case of testing the same coatings under contaminated oil lubrication conditions, the carbon coatings appeared to mitigate the pitting rather than accelerate it [39]. The mine dust, which consists of carbon and silica and has a maximum particle size of 70 μm, was utilised in the study. Despite the presence of oil contamination, a higher pitting resistance was observed for the steel–coating tribosystem than for the uncoated contact lubricated with the oil devoid of contamination.

A device incorporating a roller that operates against a disc tribosystem was utilised to analyse the wear of DLC-coated steel rollers operating with highly contaminated lubrication [53]. The schematic of the tribosystem is presented in Fig. 6.22. The tests were conducted under lubricated pure rolling conditions, with a rotational speed of 150 rpm and a load of 145 N. The coatings under investigation were deposited exclusively on the roller, which was driven by the driving disc. The roller was immersed in an oil bath contaminated with varying quantities of SiC particles, with an average size ranging from 53 to 76 μm.

Following the investigation, the wear of rollers was analysed through the assessment of weight loss and surface roughness. It was observed that the severity of wear for coated rollers was particularly pronounced, especially under conditions involving relatively high concentrations of SiC particles. The transfer of third-body abrasive particles into the contact zone resulted in the coating surface being scored more coarsely, which in turn led to further damage to the cooperating disc. It was observed that the disc which was subjected to the action of the coated rollers sustained greater levels of damage than the disc which was in contact with an uncoated roller under conditions which were identical in all respects. It should be noted that the primary

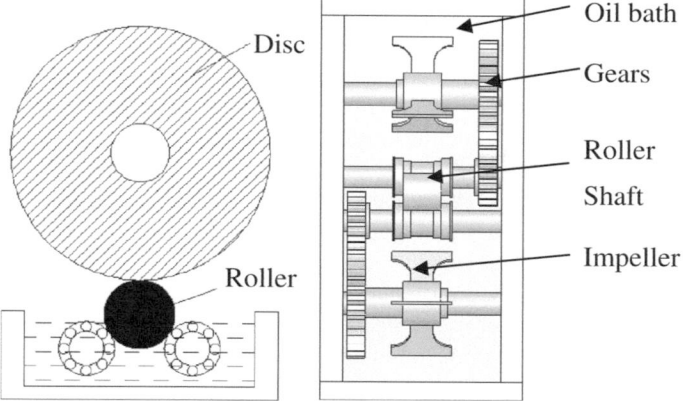

Fig. 6.22 Schematic diagram showing the specimen disc and the roller in contact. Reprinted with permission from [53]

focus of the tests was the analysis of the abrasive effect of contaminated oil on carbon coatings surfaces, rather than the determination of surface contact fatigue. However, given the potential applications of the apparatus in micropitting testing, the results of this study are relevant and merit presentation in this section.

6.2.10 *A Device with a Pair of Spur Gears*

As previously stated, the optimal methodology for evaluating tribological properties is through the utilisation of authentic specimens, with test parameters aligned with the actual operational conditions of the component or device in question. The distinctive combination of tribological properties exhibited by carbon coatings, confirmed by a substantial body of literature, renders them a promising option for enhancing the performance of gears, which are a pivotal component in advanced machinery and equipment across various sectors, including aerospace, civil and military vehicles, wind power installations, and more. The enhanced power transmission of these devices, coupled with the necessity for high reliability and failure-free performance, gives rise to the imperative of developing rigorous standards for gear performance, operating conditions and fatigue strength. A pertinent illustration of this imperative is the adaptation of a pair of spur gears test rig for the assessment of gear, engine and machine oils, as well as hydraulic oils and semi-fluid lubricants, due to galling and pitting failure [54].

Figure 6.23 shows a schematic diagram of the test rig and a view of the gear pair under test mounted in the rig. The tests are conducted in a lubricated environment at a constant speed, a constant time, a fixed and controlled lubricant temperature and a constant load (which, depending on the method, may be gradually increased with each successive test cycle). The pitting test is carried out until damage is detected, at which point it is visually assessed by estimating the area of pitting on the most damaged tooth of the small wheel. The total number of fatigue cycles is the sum of the cycles in all test runs from the beginning of the test until a pitting defect of a specified area is identified. The 50% surface fatigue life, defined as the number of fatigue cycles corresponding to a 50% probability of failure, is determined from the total number of fatigue cycles using a Weibull distribution. The second method involves subjecting the gear pair to progressively increasing loads until galling occurs on the active surface of the tooth. The evaluation of the wear is conducted through a combination of visual inspection and the utilisation of a precision balance with a measuring accuracy of 0.0001 g. The wear resistance tests are executed with a gradually increasing load, and the evaluation of the wear is undertaken in accordance with the guidelines stipulated in the standard. Figure 6.24 provides a visual representation of surface and subsurface damage to gears tested with the T12U device.

Both the visual method and the weighing method are comparative techniques. The visual assessment is inherently subjective. Furthermore, the high cost of manufacturing gears and the complicated and time-consuming nature of the test procedure mean that this type of test is not widely used. Two tests can be performed on a pair of gears (left and right tooth surfaces).

It is important to note that a significant advantage of the spur gear pair test rig is the high reliability of the test results obtained due to the frictional contact itself and the wide range of test parameters that can be selected to most accurately reflect actual operating conditions. In the context of fatigue testing of carbon coatings on a pair of spur gears, the evaluation of test results can be conducted through the utilisation of the visual method in conjunction with the Palmgren–Miner linear cumulative damage hypothesis. This approach encompasses the consideration of damage accumulation experienced by the specimen during the loading process. In instances where visual

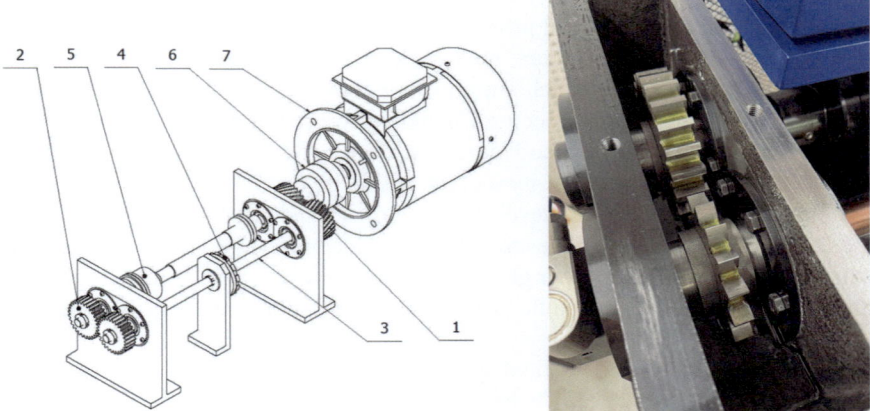

Fig. 6.23 Schematic of the T12-U device: (1) Main gear, (2) Tested gear, (3) Shaft, (4) Clutch with adjustable torque, (5) Safety clutch, (6) Engine clutch, (7) Engine

6.2 Tribotesters

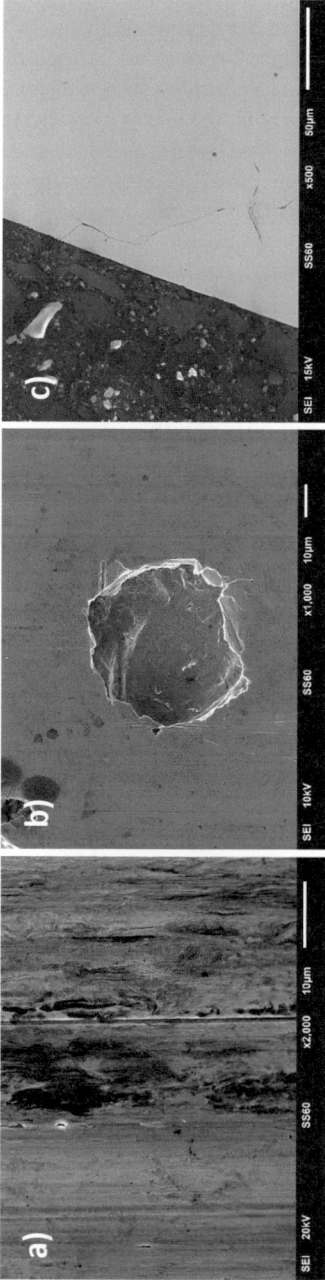

Fig. 6.24 SEM view of: **a** scuffing, **b** micropitting and **c** subsurface cracking of 3D printed AISI316L spur gears tested on T12U tribometer

analysis reveals an area of pitting greater than 4% on a gear tooth surface or evidence of spalling, the gear is deemed to have undergone fatigue failure. Conversely, scuffing resistance tests can be conducted in accordance with the FZG A/8.3/90 procedure, wherein the load on the tested pair of spur gears is incrementally increased with each subsequent test, while the surface condition is subject to a visual assessment. An additional reference method may be the analytical determination of the scuffing limit temperature, representing the maximum contact temperature at maximum load conditions. The flash temperature method is used to determine the maximum contact temperature on the tooth surface [55].

Michalczewski et al. [56] have demonstrated that the resistance of gears coated with carbon-based coatings to pitting is contingent not only on the type of coating employed, but also on the specific gear with which the coating is applied. The modification of both gears has been shown to be highly effective in addressing pitting issues. However, it has been observed that applying the coating to the gear with the higher number of teeth can enhance the fatigue life of the gears. This phenomenon can be attributed to the fact that the irregularities present on the surface of the pinion teeth are effectively eliminated through the process of polishing, a consequence of the application of a hard and wear-resistant coating. Initially, these peaks and valleys function as stress concentrators, leading to the initiation of microcracks. Additionally, the wheel with a higher number of teeth exhibits enhanced efficacy in transferring solid lubricant to the teeth of the uncoated pinion, resulting in reduced wear and improved polishing performance. Conversely, research conducted by Moorthy et al. [57] has demonstrated that micro-pits can be formed through the initiation and progression of micro-cracks within notch-like micro-valleys, resulting from the final grinding of gears. These craters are oriented in a manner that is favourably aligned with the sliding direction. The elimination of such surface irregularities, whether by polishing or by means of filling with a suitable coating, has been shown to significantly mitigate the severity of micropitting damage in gears.

The policy of sustainable development, increasing environmental awareness and new regulations regarding the use of environmentally friendly lubricants with zero or significantly reduced environmental impact are key factors in the further development of transmissions with gears and bearings working in rolling-sliding contact. The introduction of novel lubricants that are biodegradable, bio-based, and environmentally friendly has compelled manufacturers of heavily loaded machine elements to redesign their products. This redesign involves the incorporation of additional surface modifications to enhance resistance to the new environmental requirements, including increased wear, hydrogen embrittlement, and stress cracking corrosion. These factors can potentially synergistically overlap with contact fatigue stress [47, 56, 58, 59]. This, in turn, may open up new possibilities for this kind of rather expensive and time-consuming test methods, both in basic and application research, the use of which is currently marginal.

6.2.11 Other Custom Made Tribometers

The final category of devices encompasses test rigs that are meticulously engineered to analyse a specific surface-modified product, often serving as a precise replica of actual operating conditions. Notwithstanding the substantial expense and protracted testing processes, these studies represent the culminating phase of the optimisation procedures and pre-implementation research of technologic systems with considerable application potential.

Figure 6.25 illustrates the test rig employed for the examination of piston ring-cylinder liner contact. The device enables the analysis of the influence of surface roughness of carbon-based coatings on their tribological performance under hydrodynamic and mixed/boundary lubrication regimes [60].

A heavy oil and sand environment is a normal operating condition for the sucker rod pump, which is an important machine in oil exploration and extraction. A commercially available oil-well tubing was modified with DLC coating and tested under simulated working conditions for abrasion resistance (see Fig. 6.26). The findings of the tests conducted under simulated and practical working conditions have

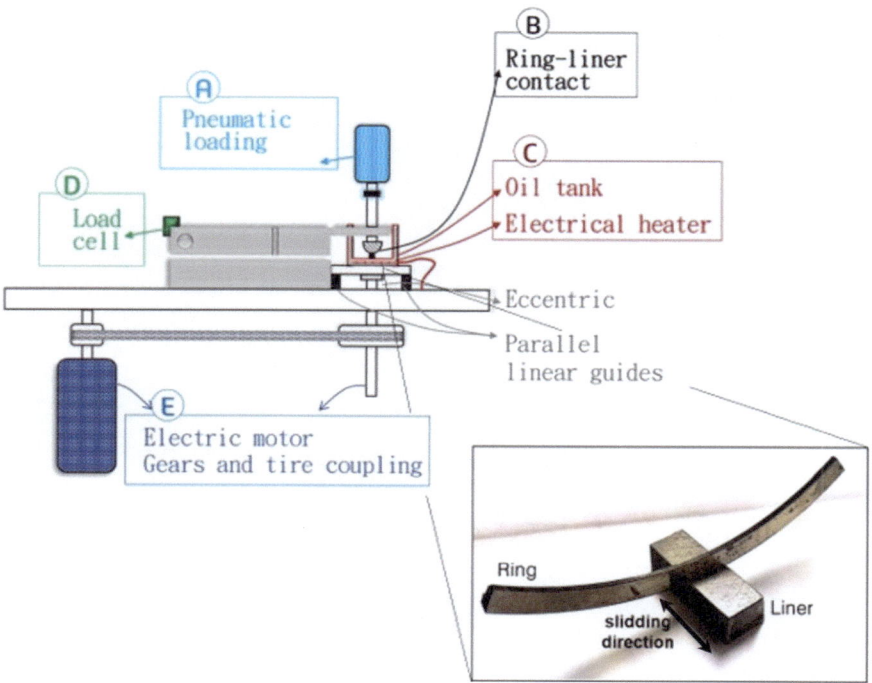

Fig. 6.25 Schematic representation of the developed test rig identifying the constituent systems and focusing the piston ring-cylinder liner samples contact. Reprinted with permission from [60]

indicated that the service life of the oil-well tubings may be enhanced by a factor of five [61].

The human body is an extremely harsh working environment in which many materials undergo corrosion, fatigue and wear, and in which all forms of surface damage interact simultaneously to lead to failure. Therefore, it is essential that every new material or new surface modification technology is validated, especially due to the serious adverse effects that can result, such as inflammation, allergy or complete rejection of the implant, which may affect a patient after hip replacement surgery. Carbon-based coatings, in general, are biocompatible materials with high corrosion resistance, but also provide low friction and wear rates, which makes them excellent candidates for modification of hip joint prosthesis heads. However, this highly responsible and demanding application requires reliable and repeatable test trials conducted in special conditions that reflect the actual operating parameters to the best extent possible. Hip joint simulators are capable of replicating wear conditions

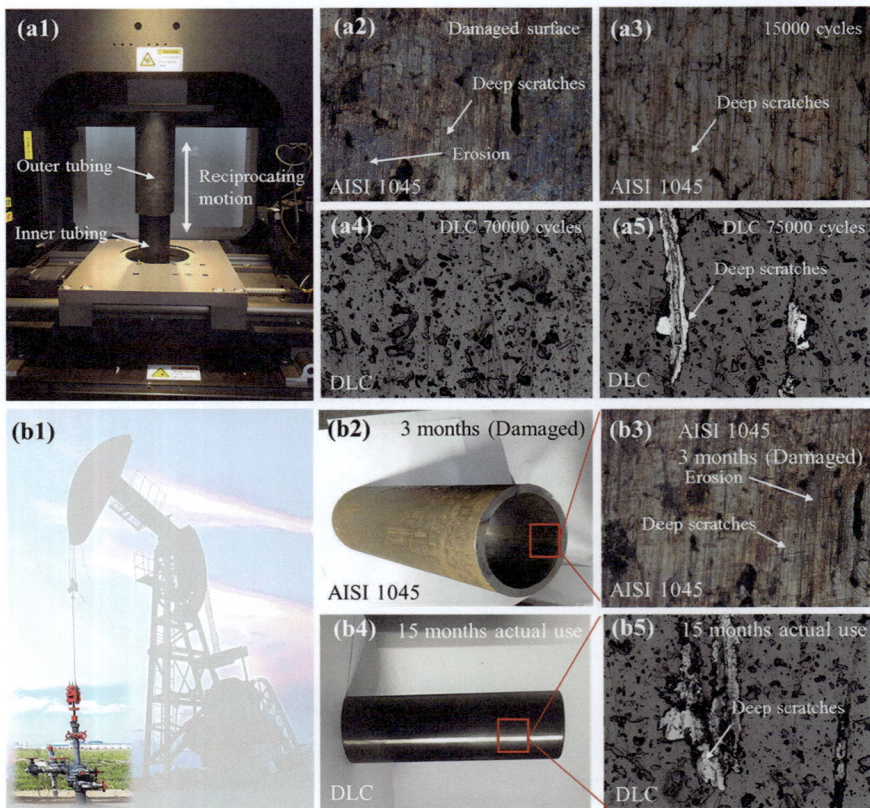

Fig. 6.26 Results of oil-well tubings on the test machine: **a1–a5** Simulated conditions; **b1–b5** Actual use. Reprinted with permission from [61]

6.2 Tribotesters

Fig. 6.27 Hip simulator. Reprinted with permission from [62]

similar to those appearing in the human body, including loads, speeds, environment and multi-directional concave shape carbon modified hip joint (see Fig. 6.27) [62]. Other literature reports indicate further technological progress in the construction of test rigs, additionally enabling a measurement of the friction coefficient of DLC modified hip joint tribological systems [63].

The preceding presentation of a variety of tribological testing techniques, encompassing both laboratory-based methods and those that accurately replicate the operational conditions of a given element or device, positions carbon-based coatings as one of the most thoroughly characterised groups of materials. A significant heterogeneity is observed in the reported values of the coefficient of friction, resistance to wear, micro-pitting, as well as nano and micro-impact, resulting from differing test conditions, the construction of test rigs and the subjective approach of laboratory personnel. The author has elected to omit a discussion of these variations, focusing instead on the presentation of the measurement principle itself and the obtained results, particularly in the case of popular and most commonly used methods. The selection of research methods and the presentation of examples were primarily guided by the distinctive tribological properties of carbon coatings, particularly in the context of sliding-rolling contact, which are distinct from those of other low-friction coating materials. However, there exist alternative tribological investigation methods that analyse the interaction between two or more bodies in relative motion, such as cavitation, solid particle, and water droplet erosion. In addition to their self-lubricating properties, carbon-based coatings offer high hardness and corrosion and wear resistance, which are particularly desirable in protective applications. These include the prevention of severe wear of rotor blades, wind turbines, and jet engines, as well as protection against hard particulate matter entrained by a fluid flow and impacting exposed surfaces [64–66]. During this dynamic process, the particle kinetic energy is dissipated through a variety of mechanisms, including ductility, fracture, heating, and phase transformations. These mechanisms are contingent on the particle (density, shape, and mechanical properties), eroded surface (mechanical properties), and operational properties (velocity and angle of incidence) [67]. However, these issues extend far beyond the scope of this book, and as such, they have been omitted from further detailed discussion.

References

1. Szczerek,. M, Wiśniewski, M.: Tribologia i tribotechnika. SIMP-PTT-ITeE, Radom, Radom (2000)
2. Capanidis, D.: Selected aspects of the methodology of tribological investiga-tions of polymer materials (2007)
3. Liu, Y., Erdemir, A., Meletis, E.I.: An investigation of the relationship between graphitization and frictional behavior of DLC coatings. Surf. Coat. Technol. **86–87**, 564–568 (1996). https://doi.org/10.1016/S0257-8972(96)03057-5
4. Ronkainen, H., Koskinen, J., Varjus, S., Holmberg, K.: Load-carrying capacity evaluation of coating/substrate systems for hydrogen-free and hydrogenated diamond-like carbon films. Tribol. Lett. **6**, 63–73 (1999). https://doi.org/10.1023/A:1019107622768
5. Zhang, T.F., Wan, Z.X., Ding, J.C., Zhang, S., Wang, Q.M., Kim, K.H.: Microstructure and high-temperature tribological properties of Si-doped hydrogenated diamond-like carbon films. Appl. Surf. Sci. **435**, 963–973 (2018). https://doi.org/10.1016/j.apsusc.2017.11.194
6. Li, W., Fan, X., Li, H., Zhu, M., Wang, L.: Probing carbon-based composite coatings toward high vacuum lubrication application. Tribol. Int. **128**, 386–396 (2018). https://doi.org/10.1016/j.triboint.2018.07.043
7. Liu, Y., Yu, B., Cao, Z., Shi, P., Zhou, N., Zhang, B., Zhang, J., Qian, L.: Probing superlubricity stability of hydrogenated diamond-like carbon film by varying sliding velocity. Appl. Surf. Sci. **439**, 976–982 (2018). https://doi.org/10.1016/j.apsusc.2018.01.048
8. Bellón Vallinot, I., de la Guerra Ochoa, E., Echávarri Otero, J., Chacón Tanarro, E., Fernández Martínez, I., Santiago Varela, J.A.: Individual and combined effects of introducing DLC coating and textured surfaces in lubricated contacts. Tribol. Int. **151** (2020). https://doi.org/10.1016/j.triboint.2020.106440
9. Jedrzejczak, A., Szymanski, W., Kolodziejczyk, L., Sobczyk-Guzenda, A., Kaczorowski, W., Grabarczyk, J., Niedzielski, P., Kolodziejczyk, A., Batory, D.: Tribological characteristics of a-C:H:Si and a-C:H:SiOx coatings tested in simulated body fluid and protein environment. Materials **15** (2022). https://doi.org/10.3390/ma15062082
10. Statuti, R.P.C.C., Radi, P.A., Santos, L.V., Trava-Airoldi, V.J.: A tribological study of the hybrid lubrication of DLC films with oil and water. Wear **267**, 1208–1213 (2009). https://doi.org/10.1016/j.wear.2008.11.033
11. Hinüber, C., Kleemann, C., Friederichs, R.J., Haubold, L., Scheibe, H.J., Schuelke, T., Boehlert, C., Baumann, M.J.: Biocompatibility and mechanical properties of diamond-like coatings on cobalt-chromium-molybdenum steel and titanium-aluminum-vanadium biomedical alloys. J. Biomed. Mater. Res. A **95A**, 388–400 (2010). https://doi.org/10.1002/jbm.a.32851
12. Carapeto, A.P., Serro, A.P., Nunes, B.M.F., Martins, M.C.L., Todorovic, S., Duarte, M.T., André, V., Colaço, R., Saramago, B.: Characterization of two DLC coatings for joint prosthesis: the role of albumin on the tribological behavior. Surf. Coat. Technol. **204**, 3451–3458 (2010). https://doi.org/10.1016/j.surfcoat.2010.04.022

References

13. Bystrzycka, E., Prowizor, M., Piwoński, I., Kisielewska, A., Batory, D., Jędrzejczak, A., Dudek, M., Kozłowski, W., Cichomski, M.: The effect of fluoroalkylsilanes on tribological properties and wettability of Si-DLC coatings. Mater. Res. Express **5**, 036411 (2018). https://doi.org/10.1088/2053-1591/aab472
14. Bonnevie, E.D., Baro, V.J., Wang, L., Burris, D.L.: In situ studies of cartilage microtribology: roles of speed and contact area. Tribol. Lett. **41**, 83–95 (2011). https://doi.org/10.1007/s11249-010-9687-0
15. Kubik, A., Hadryś, D., Stanik, Z., Jasiok, M.: Analysis of tribological wear in block—on ring contact on tribological tester T-05. Sci. J. Silesian Univ. Technol. Ser. Transp. **105**, 113–123 (2019). https://doi.org/10.20858/sjsutst.2019.105.10
16. Haque, T., Ertas, D., Ozekcin, A., Jin, H.W., Srinivasan, R.: The role of abrasive particle size on the wear of diamond-like carbon coatings. Wear **302**, 882–889 (2013). https://doi.org/10.1016/j.wear.2013.01.080
17. Nobili, L., Magagnin, L.: DLC coatings for hydraulic applications. Trans. Nonferrous Metals Soc. China **19**, 810–813 (2009). https://doi.org/10.1016/S1003-6326(08)60355-6
18. Grierson, D.S., Carpick, R.W.: Nanotribology of carbon-based materials. Nano Today **2**, 12–21 (2007). https://doi.org/10.1016/S1748-0132(07)70139-1
19. Enachescu, M., Van Den Oetelaar, R.J.A., Carpick, R.W., Ogletree, D.F., Flipse, C.F.J., Salmeron, M.: Observation of proportionality between friction and contact area at the nanometer scale (1999)
20. Shi, P., Sun, J., Yan, W., Zhou, N., Zhang, J., Zhang, J., Chen, L., Qian, L.: Roles of phase transition and surface property evolution in nanotribological behaviors of H-DLC: Effects of thermal and UV irradiation treatments. Appl. Surf. Sci. **514** (2020). https://doi.org/10.1016/j.apsusc.2020.145960
21. Mann, A.B.: Nanotribology and Nanomechanics. Springer International Publishing, Cham (2017)
22. Bhushan, B., Israelachvili, J.N., Landman, U.: Nanotribology: friction, wear and lubrication at the atomic scale. Nature **374**, 607–616 (1995). https://doi.org/10.1038/374607a0
23. Buzio, R., Boragno, C., Valbusa, U.: Nanotribology of cluster assembled carbon films. Wear **254**, 981–987 (2003). https://doi.org/10.1016/S0043-1648(03)00303-X
24. Wang, M., Miyake, S., Matsunuma, S.: Nanowear studies of PFPE lubricant on magnetic perpendicular recording DLC-film-coated disk by lateral oscillation test. Wear **259**, 1332–1342 (2005). https://doi.org/10.1016/j.wear.2005.03.026
25. Miyake, S., Yamazaki, S.: Evaluation of protuberance and groove formation in extremely thin DLC films on Si substrates due to diamond tip sliding by atomic force microscopy. Wear **318**, 135–144 (2014). https://doi.org/10.1016/j.wear.2014.06.018
26. Kvasnica, S., Schalko, J., Eisenmenger-Sittner, C., Benardi, J., Vorlaufer, G., Pauschitz, A., Roy, M.: Nanotribological study of PECVD DLC and reactively sputtered Ti containing carbon films. Diam. Relat. Mater. **15**, 1743–1752 (2006). https://doi.org/10.1016/j.diamond.2006.03.005
27. Singh, R.A., Na, K., Yi, J.W., Lee, K.R., Yoon, E.S.: DLC nano-dot surfaces for tribological applications in MEMS devices. Appl. Surf. Sci. **257**, 3153–3157 (2011). https://doi.org/10.1016/j.apsusc.2010.10.131
28. Kolodziejczyk, L., Szymanski, W., Batory, D., Jedrzejczak, A.: Nanotribology of silver and silicon doped carbon coatings. Diam. Relat. Mater. **67** (2016). https://doi.org/10.1016/j.diamond.2015.12.010
29. Beake, B.D., Goodes, S.R., Smith, J.F., Madani, R., Rego, C.A., Cherry, R.I., Wagner, T.: Investigating the fracture resistance and adhesion of DLC films with micro-impact testing (2002)
30. Rueda-Ruiz, M., Beake, B.D., Molina-Aldareguia, J.M.: New instrumentation and analysis methodology for nano-impact testing. Mater. Des. **192** (2020). https://doi.org/10.1016/j.matdes.2020.108715
31. Bouzakis, K.D., Siganos, A., Leyendecker, T., Erkens, G.: Thin hard coatings fracture propagation during the impact test. Thin Solid Films **460**, 181–189 (2004). https://doi.org/10.1016/j.tsf.2004.02.009

32. Abdollah, M.F., Bin, Yamaguchi, Y., Akao, T., Inayoshi, N., Tokoroyama, T., Umehara, N.: The effect of maximum normal impact load, absorbed energy, and contact impulse, on the impact crater volume/depth of DLC coating. Tribolo. Online **6**, 257–264 (2011). https://doi.org/10.2474/trol.6.257
33. McMaster, S.J., Liskiewicz, T.W., Neville, A., Beake, B.D.: Probing fatigue resistance in multilayer DLC coatings by micro- and nano-impact: correlation to erosion tests. Surf. Coat. Technol. **402** (2020). https://doi.org/10.1016/j.surfcoat.2020.126319
34. Beake, B.D.: Evaluation of the fracture resistance of DLC coatings on tool steel under dynamic loading. Surf. Coat. Technol. **198**, 90–93 (2005). https://doi.org/10.1016/j.surfcoat.2004.10.048
35. Dorner, A., Schürer, C., Reisel, G., Irmer, G., Seidel, O., Müller, E.: Diamond-like carbon-coated Ti–6A–l4V: influence of the coating thickness on the structure and the abrasive wear resistance. Wear **249**, 489–497 (2001)
36. Hainsworth, S.V., Uhure, N.J.: Diamond like carbon coatings for tribology: Production techniques, characterisation methods and applications (2007)
37. Michler, T., Siebert, C.: Abrasive wear testing of DLC coatings deposited on plane and cylindrical parts. Surf. Coat. Technol. **163**, 546–551 (2003). https://doi.org/10.1016/S0257-8972(02)00620-5
38. Gbhlin, R., Larsson, M., Jacobson, S., Hogmark, S.: The crater grinder method as a means for coating wear evaluation-an update (1997)
39. Tuszyński, W., Michalczewski, R., Osuch-Słomka, E., Snarski-Adamski, A., Kalbarczyk, M., Wieczorek, A.N., Nędza, J.: Abrasive wear, scuffing and rolling contact fatigue of DLC-coated 18CrNiMo7-6 steel lubricated by a pure and contaminated gear oil. Materials **14** (2021). https://doi.org/10.3390/ma14227086
40. Batory, D., Szymanski, W., Clapa, M.: Mechanical and tribological properties of gradient a-C:H/Ti coatings. Mater. Sci. Poland **31**, 415–423 (2013). https://doi.org/10.2478/s13536-013-0121-9
41. Rosado, L., Jain, V.K., Trivedi, H.K.: The effect of diamond-like carbon coatings on the rolling fatigue and wear of M50 steel. Wear **212**, 1–6 (1997). https://doi.org/10.1016/S0043-1648(97)00147-6
42. Stewart, S., Ahmed, R.: Rolling contact fatigue of surface coatings-a review (2002)
43. Fu, H., Rivera-Díaz-del-Castillo, P.E.J.: A unified theory for microstructural alterations in bearing steels under rolling contact fatigue. Acta Mater. **155**, 43–55 (2018). https://doi.org/10.1016/j.actamat.2018.05.056
44. Wei, R., Wilbur, P.J., Liston, M.-J., Lux, G.: Rolling-contact-fatigue wear characteristics of diamond-like hydrocarbon coatings on steels. Wear **162–164**, 558–568 (1993). https://doi.org/10.1016/0043-1648(93)90541-S
45. Morales-Espejel, G.E., Brizmer, V.: Micropitting modelling in rolling-sliding contacts: application to rolling bearings. Tribol. Trans. **54**, 625–643 (2011). https://doi.org/10.1080/10402004.2011.587633
46. Morales-Espejel, G.E., Gabelli, A.: The progression of surface rolling contact fatigue damage of rolling bearings with artificial dents. Tribol. Trans. **58**, 418–431 (2015). https://doi.org/10.1080/10402004.2014.983251
47. Zapata Tamayo, J.G., Björling, M., Shi, Y., Hardell, J., Larsson, R.: Micropitting performance and friction behaviour of DLC coated bearing steel surfaces: On the influence of Glycerol-based lubricants. Tribol. Int. **196** (2024). https://doi.org/10.1016/j.triboint.2024.109674
48. Oila, A., Bull, S.J.: Assessment of the factors influencing micropitting in rolling/sliding contacts. Wear **258**, 1510–1524 (2005). https://doi.org/10.1016/j.wear.2004.10.012
49. Weibring, M., Gondecki, L., Tenberge, P.: Simulation of fatigue failure on tooth flanks in consideration of pitting initiation and growth. Tribol. Int. **131**, 299–307 (2019). https://doi.org/10.1016/j.triboint.2018.10.029
50. Wu, J., Wang, L., He, T., Wang, T., Shu, K., Gu, L., Zhang, C.: Mixed lubrication of coated angular contact ball bearing considering dynamic characteristics. Lubr. Sci. **33**, 201–213 (2021). https://doi.org/10.1002/ls.1538

51. Singh, H., Ramirez, G., Eryilmaz, O., Greco, A., Doll, G., Erdemir, A.: Fatigue resistant carbon coatings for rolling/sliding contacts. Tribol. Int. **98**, 172–178 (2016). https://doi.org/10.1016/j.triboint.2016.02.008
52. Michalczewski, R., Kalbarczyk, M., Mańkowska-Snopczyńska, A., Osuch-Słomka, E., Piekoszewski, W., Snarski-Adamski, A., Szczerek, M., Tuszyński, W., Wulczyński, J., Wieczorek, A.: The effect of a gear oil on abrasion, scuffing, and pitting of the DLC-coated 18CrNiMo7-6 Steel. Coatings **9**, 2 (2018). https://doi.org/10.3390/coatings9010002
53. He, F., Wong, P.L., Zhou, X.: Wear properties of DLC-coated steel rollers running with highly contaminated lubrication. Tribol. Int. **43**, 990–996 (2010). https://doi.org/10.1016/j.triboint.2009.12.058
54. Martins, R., Amaro, R., Seabra, J.: Influence of low friction coatings on the scuffing load capacity and efficiency of gears. Tribol. Int. **41**, 234–243 (2008). https://doi.org/10.1016/j.triboint.2007.05.008
55. Wu, J., Wei, P., Liu, G., Chen, D., Zhang, X., Chen, T., Liu, H.: A comprehensive evaluation of DLC coating on gear bending fatigue, contact fatigue, and scuffing performance. Wear 536–537 (2024). https://doi.org/10.1016/j.wear.2023.205177
56. Michalczewski, R., Kalbarczyk, M., Piekoszewski, W., Szczerek, M., Tuszyński, W.: The rolling contact fatigue of WC/C-coated spur gears. Proc. Inst. Mech. Eng. Part J: J. Eng. Tribol. **227**, 850–860 (2013). https://doi.org/10.1177/1350650113478179
57. Moorty V., Shaw B.: An observation on the initiation of micro-pitting damage in as-ground and coated gears during contact fatigue. Wear **297**, 878–884 (2013). https://doi.org/10.1016/j.wear.2012.11.001
58. Martin, J.M., De Barros-Bouchet, M.I.: Water-like lubrication of hard contacts by polyhydric alcohols. In: Aqueous Lubrication, pp. 219–235. Co-Published with Indian Institute of Science (IISc), Bangalore, India (2014)
59. Singh, Y., Farooq, A., Raza, A., Mahmood, M.A., Jain, S.: Sustainability of a non-edible vegetable oil based bio-lubricant for automotive applications: a review. Process. Saf. Environ. Prot. **111**, 701–713 (2017). https://doi.org/10.1016/j.psep.2017.08.041
60. Ferreira, R., Almeida, R., Carvalho, Ó., Sobral, L., Carvalho, S., Silva, F.: Influence of a DLC coating topography in the piston ring/cylinder liner tribological performance. J. Manuf. Process. **66**, 483–493 (2021). https://doi.org/10.1016/j.jmapro.2021.04.044
61. Liu, L., Wu, Z., Cui, S., An, X., Ma, Z., Shao, T., Fu, R.K.Y., Wang, R., Lin, H., Pan, F., Chu, P.K.: Abrasion and erosion behavior of DLC-coated oil-well tubings in a heavy oil/sand environment. Surf. Coat. Technol. **357**, 379–383 (2019). https://doi.org/10.1016/j.surfcoat.2018.09.081
62. Liu, H., Leng, Y., Tang, J., Wang, S., Xie, D., Sun, H., Huang, N.: Tribological performance of ultra-high-molecular-weight polyethylene sliding against DLC-coated and nitrogen ion implanted CoCrMo alloy measured in a hip joint simulator. Surf. Coat. Technol. **206**, 4907–4914 (2012). https://doi.org/10.1016/j.surfcoat.2012.05.090
63. Choudhury, D., Urban, F., Vrbka, M., Hartl, M., Krupka, I.: A novel tribological study on DLC-coated micro-dimpled orthopedics implant interface. J. Mech. Behav. Biomed. Mater. **45**, 121–131 (2015). https://doi.org/10.1016/j.jmbbm.2014.11.028
64. Mednikov, A., Ryzhenkov, A., Zilova, O., Tkhabisimov, A., Kachalin, G., Sidorov, S.: Study of stress state changes in steel with Ti–TiC-DLC coating under high speed droplet impact. Tribol. Int. **173** (2022). https://doi.org/10.1016/j.triboint.2022.107626
65. Wu, S., Chen, S., Zhang, L., Yin, Z., Liu, Y., Qi, Y., Liao, B., Zhang, X., Ouyang, X., Chen, L., Wang, J.: Study on preparation and anti-sand erosion performance of thick DLC coating combined with FCVA deposition and HVP technology. Vacuum **229** (2024). https://doi.org/10.1016/j.vacuum.2024.113532
66. Depner-Miller, U., Ellermeier, J., Scheerer, H., Oechsner, M., Bobzin, K., Bagcivan, N., Brögelmann, T., Weiss, R., Durst, K., Schmid, C.: Influence of application technology on the erosion resistance of DLC coatings. Surf. Coat. Technol. **237**, 284–291 (2013). https://doi.org/10.1016/j.surfcoat.2013.07.043
67. Bousser, E., Martinu, L., Klemberg-Sapieha, J.E.: Solid particle erosion mechanisms of protective coatings for aerospace applications. Surf. Coat. Technol. **257**, 165–181 (2014). https://doi.org/10.1016/j.surfcoat.2014.08.037

Chapter 7
Methods of Characterization of Carbon Based Low Friction Coatings After Tribological Testing

Abstract The chapter presents a range of research methods used in the surface examination of carbon coatings after tribological testing. The fundamental investigation techniques and apparatus employed in the post-mortem analysis of the profile, chemical structure and composition of wear tracks and wear scars on cooperating surfaces are presented. The general approach, data treatment and examples of results presentation are given. Some advantages and disadvantages of particular testing methods are discussed.

It is evident that the outcomes of friction and wear testing of diamond-like carbon layers serve as a substantial repository of knowledge pertaining to their properties and operational parameters. The wear rate and coefficient of friction are widely regarded as the fundamental tribological characteristics of DLC and other low-friction materials. These characteristics are incorporated into subsequent database cells, which are indispensable for the processes of selection and optimisation of surface modification of functional materials for a specific application. Conversely, the value of linear wear and the coefficient of friction mark the commencement of the actual research process, aimed at comprehending and explaining particular frictional behaviours or at comparing the obtained results with other scientific reports. In this chapter, the fundamental investigation techniques and apparatus employed in the post-mortem analysis of the profile, chemical structure and composition of wear tracks and wear scars on cooperating surfaces will be presented. The general approach and data treatment will be of particular interest, without unnecessary elaboration on the principles of operation, accuracy, or the superiority of one technique over another.

7.1 Profilometry

The fundamental technique that facilitates the quantitative analysis of surface wear in the friction process is profilometry, which can be performed using either a stylus or an optical profilometer, depending on the available option. The measurement process entails the analysis of the topography of the abrasion mark, including its shape, depth, and width. Depending on the technique employed, the measurement with a stylus profilometer is two-dimensional, necessitating the acquisition of data at multiple points along the wear track and the subsequent calculation of the mean value. The wear volume is then calculated by multiplying the surface area of the projection of the abrasion onto the cross-section plane by its length. More advanced optical profilometers are capable of scanning the entire wear track on the sample and the wear scar on the counterbody in three dimensions. The spatial profile that has been registered can then be analysed using a variety of surface analysis software for profilometers. This analysis allows the study of wear in every point of the wear track and the wear scar, together with its qualitative characteristics. As previously described, many tribological testing techniques are normalised, enabling direct comparison of the results with other scientific reports. The ball-on-disc test is the most popular method, but not the only one. A common practice in this case is to determine an absolute wear rate value independent of the applied load or the specified sliding distance. For that purpose, the following formula can be applied, either for the coating or the counterbody:

$$W_r = \frac{W_{vol}[mm^3]}{L[N] * d[m]} \qquad (7.1)$$

where:

W_r wear rate
W_{vol} total volume of the wear
L load
d sliding distance

Figure 7.1 illustrates an example of a 3D scan of an a wear-track recorded using the interferometer, while Fig. 7.2 presents a cross-sectional view, both obtained by using 3D image analysis software.

Fig. 7.1 An example of a 3D scan of a wear-track

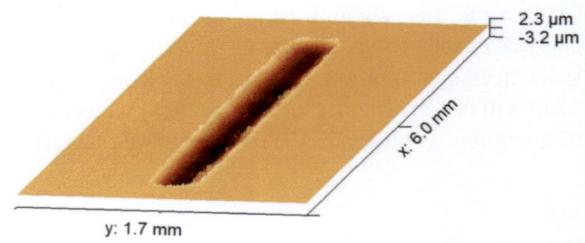

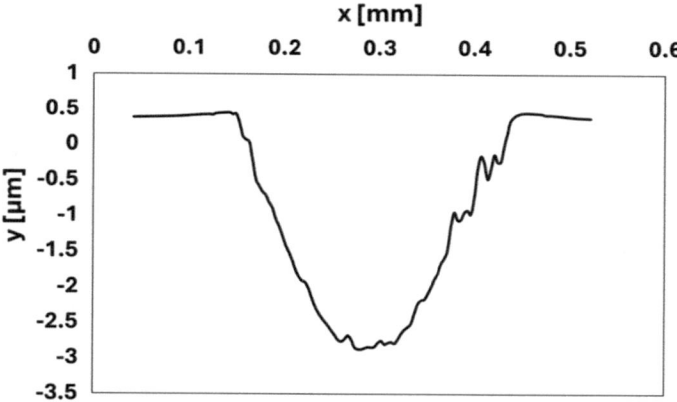

Fig. 7.2 An example of cross-section view of a wear-track

7.2 Optical and Scanning Electron Microscopy

Structure and morphology analysis is typically performed using optical or scanning electron microscopy. In most cases, both the wear track and the wear scar can be analysed. At first glance, it is possible to determine the main geometrical parameters, namely the width of the wear track and the dimensions of the wear scar (for spherical counterbodies, the abrasion mark has a circular or elliptical shape). Subsequent analysis determines the nature of the wear, categorised as either abrasive, adhesive, fatigue, or indicative of a third-body effect. Scanning electron microscopy is the superior technique in this regard, offering enhanced depth of field, a broader range of useful magnifications, and the capability to analyse the chemical composition of both the wear track and the counterbody. Secondary and backscattered electron images provide information on topography and phase contrast, respectively. The latter is especially useful in the analysis of possible material transfer between the analysed surface and the counterbody. Chemical composition analysis may bring some insight into the character of material transfer, oxidation or possible chemical reactions. Figure 7.3 presents an optical image of the wear scar on ZrO_2 counterbody and a scanning electron image with the EDS analysis of the wear track on silicon incorporated DLC layer.

The optical microscopy image provides a comprehensive overview of the wear scar, facilitating straightforward measurement. However, for in-depth analysis of the wear track, the optical microscope's magnification falls short of the necessary level of detail. The electron microscope's enhanced resolution and depth of field offer a superior perspective, enabling more precise insight into the wear characteristics and the identification of coating spallation regions. A chemical composition analysis revealed the transfer of counter-sample material to the coating surface, leading to its accumulation in local defects resulting from wear processes. It is important to note that the diameter of the interaction volume between the sample surface and the

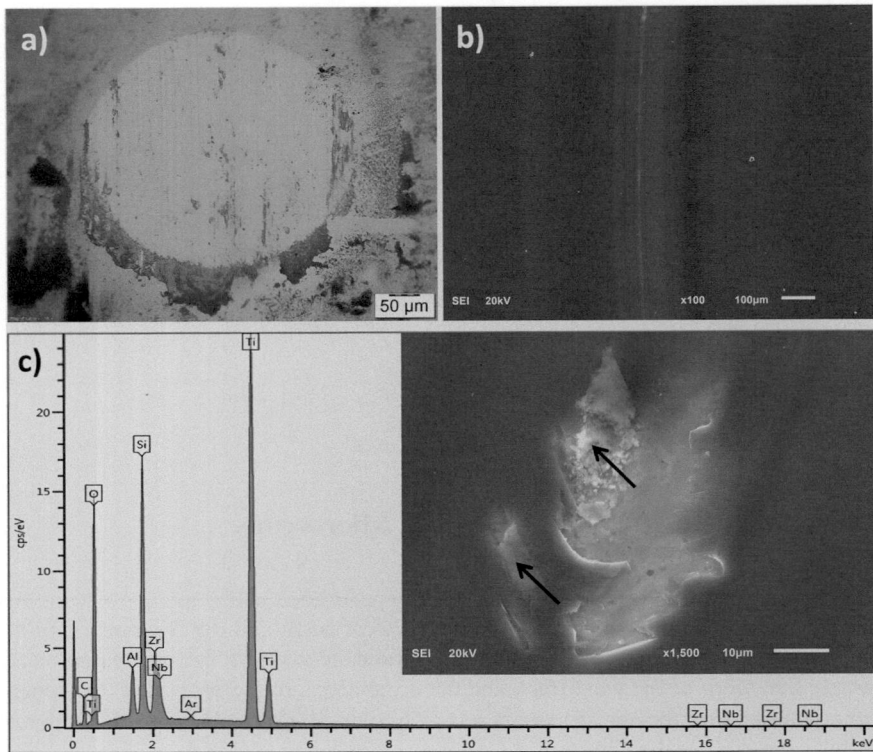

Fig. 7.3 Surface morphology analysis results: **a** optical microscope view of the ZrO$_2$ counterbody, **b** SEM image of the wear track on a-C:H:SiO$_x$ coating and **c** EDS analysis of the wear products in the wear track. Reprinted with permission from [1]

primary electron beam is larger than the electron spot. This results in the generation of characteristic X-ray radiation from a specific volume at a depth of 1–3 μm, the extent of which depends on the acceleration voltage. Consequently, without proper calibration of the EDS system, the results of the chemical composition analysis cannot be considered quantitative. Nevertheless, even in the absence of calibration, the EDS analysis provides valuable insights into the tribological process, thereby enhancing its reliability through qualitative confirmation.

The rapid development of technology has resulted in a growing trend of standard SEM units being equipped with focused ion beam (FIB) technology. FIB employs a focused beam of ions to locally remove or mill away material from a sample. This technique is capable of treatments such as pattern formation with a resolution of ~10 nm, as well as uniform surface polishing [2]. In the study of the wear tracks and wear scars, this technique is particularly useful for the site-selective preparation of samples for high-resolution imaging techniques, in particular transmission electron microscopy, as well as for 3D imaging using scanning electron microscopy [3]. In

7.3 X-Ray Photoelectron Spectroscopy

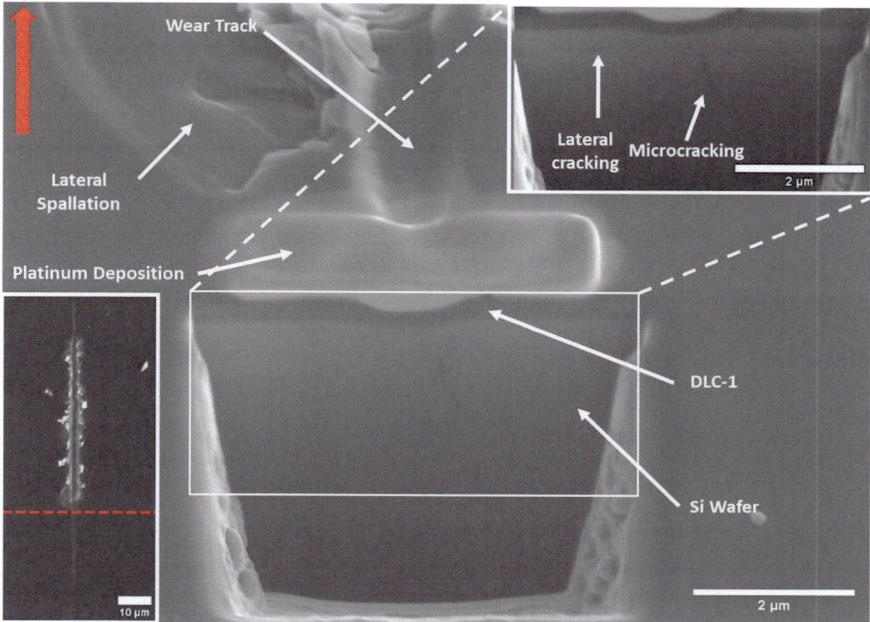

Fig. 7.4 FIB cross section of the pre spallation region of a 5 μm radius diamond ex situ scratch test of DLC coating imaged at a tilt angle of 52°. Scratch direction is indicated by the red arrow. Reprinted with permission under Creative Commons CC-BY from [4]

Fig. 7.4 is presented SEM image of cross-section of the wear track on the surface of the carbon coating prepared using FIB technique. Following the protection of the surface of the wear track with platinum coating, a volume of the material was milled by means of the FIB system in the direction normal to the surface. As a result, a good quality cross section was obtained showing the coating, the interface and the subsurface effects induced by the interaction of the sample with the diamond indenter. The motorised sample stage and the fact that FIB systems are integrated with SEMs allows the preparation of the cross section and its subsequent analysis without the need to transfer the sample from one instrument to another.

7.3 X-Ray Photoelectron Spectroscopy

X-ray Photoelectron Spectroscopy (XPS) is a surface-sensitive analytical technique that provides more accurate results of chemical composition analysis before and after tribological investigations. In this technique, X-rays are bombarded onto the surface of a material and the kinetic energy of the emitted electrons is measured. The two major advantages of this technique that make it a powerful analytical method are its surface sensitivity and its ability to reveal the chemical state information of the elements in the sample. However, it is important to note that the interpretation of data on both a qualitative and quantitative level requires a solid foundation in solid-state physics and extensive experience. Despite the increasing accessibility and user-friendliness of XPS devices, still the published data is often misinterpreted [5, 6]. A representative XPS spectra registered for a silicon-incorporated carbon coating is presented in Fig. 7.5.

The analysis of wear tracks using X-ray photoelectron spectroscopy facilitates the determination of surface chemistry, encompassing the oxidation of existing compounds and the creation of new ones, as well as the decomposition of existing

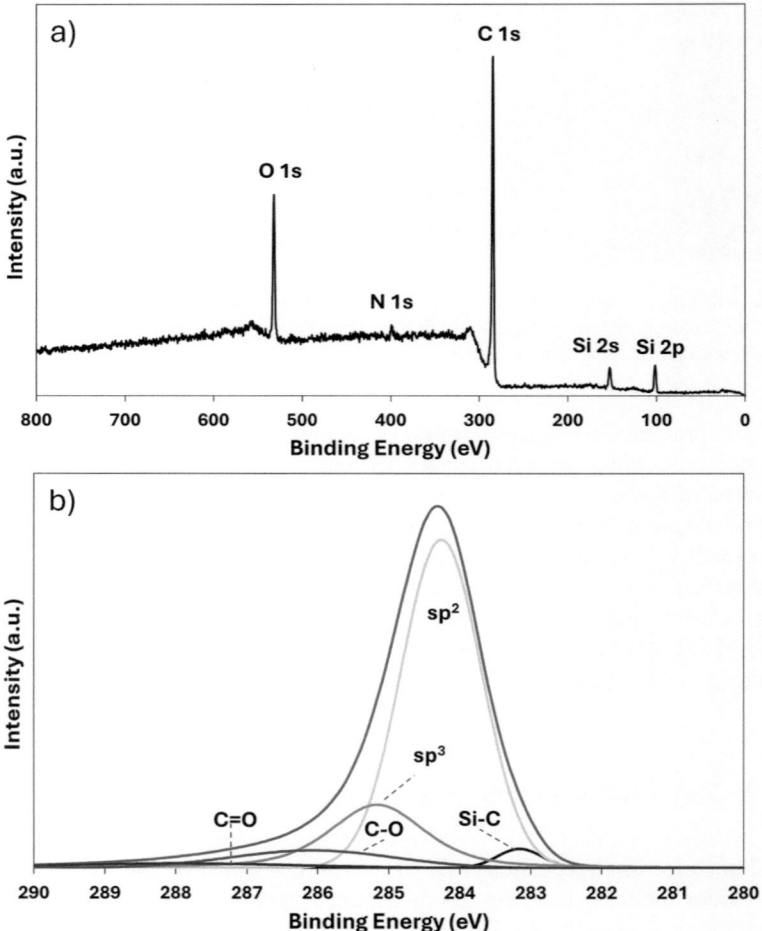

Fig. 7.5 **a** XPS survey spectra of Si—incorporated DLC film; **b** C 1 s XPS spectra fitted with five Gaussians

ones. Additionally, surface functionalisation, including the interaction of the tribological system with the surrounding environment, can be determined. Typically, XPS spectra are collected from both the wear track and from areas outside the track in order to facilitate a comparison of the results. For instance, XPS analysis of wear tracks on Gd-incorporated DLC films tested tribologically with ionic liquids explored the interaction between bromide ions and the gadolinium dopant, confirming the formation of phosphine oxide and additional P–O bonds, which led to the formation of a phosphorus-rich tribofilm that enhanced wear resistance [7]. The effect of contact between steel and DLC contact, lubricated by a fully formulated 5W30 engine oil in boundary/mixed lubrication regime was analysed using the XPS method. The study revealed that with an increase in contact load, a greater quantity of calcium carbonate-rich patchy tribofilm was formed on the steel ring [8]. In the context of high-temperature applications, the pivotal factor pertains to the interaction of the coating materials with atmospheric oxygen under specific friction conditions. Consequently, the O1s spectra recorded within the wear track frequently hold particular interest. Boron and chromium co-doped DLC coatings, tested tribologically at 300 °C, revealed that for low chromium contents in the coating, the presence of boron counteracted the oxidation processes, whereas when the Cr concentration increased it promoted the formation of Cr_2O_3, which was considered an adverse third body leading to severe wear [9]. The latter example may be confusing, since EDS analysis may yield similar results. However, oxygen contamination of the surface is a common problem, and it is difficult to distinguish between surface contamination and the actual formation of oxide as a result of a friction test. In this case, another advantage of XPS is evident. Namely, the possibility of etching the tested substrate surface layer by layer and subsequently analysing its chemical composition without direct oxygen access, thus providing the actual chemical composition [10].

As previously outlined, carbon-based coating materials possess an amorphous structure, devoid of long-range order. Consequently, the XPS technique emerges as a suitable method for determining the concentration of σsp^3 and σsp^2 hybridised carbon bonds within the coating material. Indeed, such analysis is feasible and can be utilised to ascertain the degree of graphitisation of DLC coatings and the formation of a transition layer during the tribological process. However, the most suitable technique for this purpose is Raman Spectroscopy.

7.4 Raman Spectroscopy

Raman spectroscopy is a non-destructive chemical analysis technique which provides detailed information about the chemical structure, phase and polymorphy, crystallinity, intrinsic stress/strain, as well as contamination and impurity of a substance. It is based on the inelastic scattering of light, whereby a molecule scatters incident light from a high intensity laser light source. A small amount of light is scattered at different wavelengths, which depend on the chemical structure of the analyte. The Raman spectrum thus features a number of peaks, showing the intensity and wavelength position of the Raman scattered light, with each peak corresponding to a specific molecular bond vibration, including individual bonds such as C–C, C = C, N–O, C–H etc.

In the domain of research on Raman spectroscopy of carbon coatings, the contributions of Casiraghi, Ferrari and Robertson [11–13] have been significant to its advancement. The Raman spectra of diamond-like carbon coatings for visible excitation typically exhibit a broad asymmetric peak within the range of 1400–1700 cm^{-1}. Subsequent analysis involves the fitting of the spectrum with two Gaussians, centred at frequencies of approximately 1350 and 1530 cm^{-1}, which are commonly designated as the D (disorder) and G (graphite) bands, respectively. A typical Raman spectrum deconvolution, considering these two contributions, is illustrated in Fig. 7.6.

The presence of the D band is attributed to the existence of vibrations of sp^2 carbon atoms organised only in aromatic rings (breathing mode) and the G-peak is attributed to the stretching mode vibrations of sp^2-bonded carbon atoms pairs in both chains and rings. An in-depth analysis of the particular parameters of both D and G bands may bring additional information related to carbon bonding with regard to the sp^3 component, which in fact is only a qualitative indication [15]. The number and size of sp^2 clusters can be determined based on the I_D/I_G intensity ratio, while the full-width at half maximum of the G peak (FWHM) is related to the measure of disorder in the film. Furthermore, the position of the G band provides insight into the residual stress of the coating.

There are Raman spectrometers which are equipped with a UV excitation source. In such cases, an additional peak emerges in the Raman spectra at frequency of

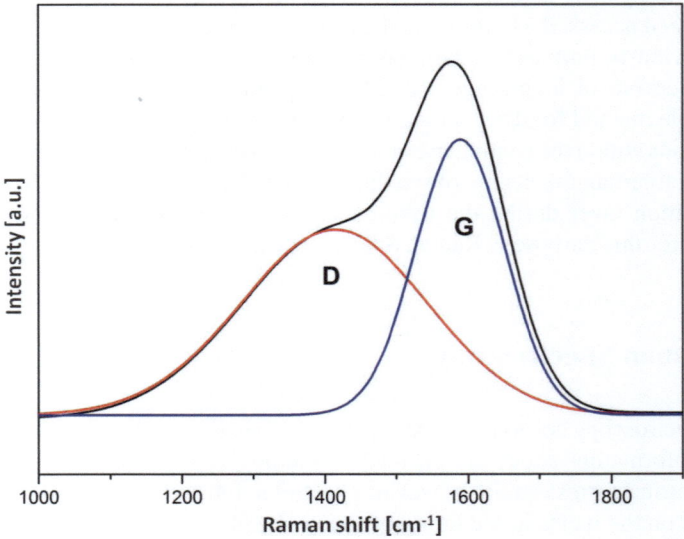

Fig. 7.6 An example of the deconvolution of a Raman spectrum of DLC coating. Reprinted with permission from [14]

7.4 Raman Spectroscopy

approximately 1060 cm^{-1} (a T peak), and this is attributable to C–C sp^3 vibrations [11]. Nevertheless, results derived from such techniques are seldom reported in the extant literature.

A comparison of Raman spectra registered for silicon and oxygen incorporated coating samples following tribological investigations involving various counterbodies is illustrated in Fig. 7.7. The results of deconvolution and key parameters of the D and G bands are presented in Table 7.1.

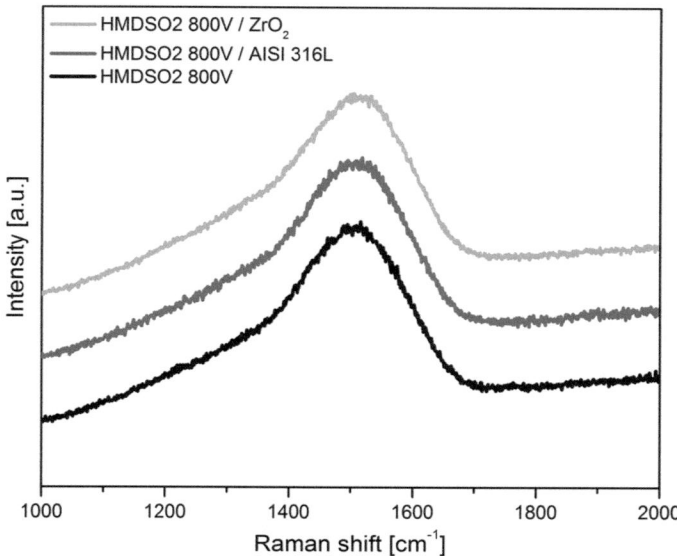

Fig. 7.7 A comparison of Raman spectra registered in the middle of the coating and in the wear tracks for AISI 316L and ZrO$_2$ counterbodies. Reprinted with permission from [1]

Table 7.1 Comparison of the results of deconvolution of Raman spectra for DLC and a-C:H:SiO$_x$ samples before and after the friction tests

Sample	Coating		Wear track AISI316L		Wear track ZrO$_2$	
	G band position [cm^{-1}]	I_D/I_G	G band position [cm^{-1}]	I_D/I_G	G band position [cm^{-1}]	I_D/I_G
DLC 600 V	1537	0.62	1543	0.82	1542	0.82
HMDSO1 600 V	1522	0.47	1526	0.50	1525	0.58
DLC 800 V	1560	1.42	1562	1.61	1564	1.65
HMDSO1 800 V	1542	0.97	1545	0.98	1545	1.11
HMDSO2 800 V	1510	0.72	1513	0.80	1512	0.87

In the case of unmodified DLC coating a slight shift of the G band into higher wavenumbers and an increased I_D/I_G ratio compared to the reference samples indicate that the coatings are undergoing graphitization as a result of friction induced transformation. A much less intense shift of the G band and increase in the I_D/I_G ratio are registered for a-C:H:SiO$_x$ coatings, suggesting that silicon prevents the graphitisation of the amorphous diamond-like carbon matrix under dry friction conditions.

The automated sample stage systems also allow conducting the Raman mapping across the wear track. An example of analysis of variation of the G-band position and full-width at half maximum across the wear track are presented in Fig. 7.8.

As demonstrated in the attached spectra, it is evident that the amorphous structure of the carbon coating in the wear track exhibits a clear trend of increasing order. In addition to this, three high-intensity peaks can be distinguished, as compared to the average value of the G-band position at approximately 1545 cm^{-1}. Consequently, it can be concluded that, in addition to the overall wear track analysis, the wear products that are still present in the wear track can be easily distinguished and examined.

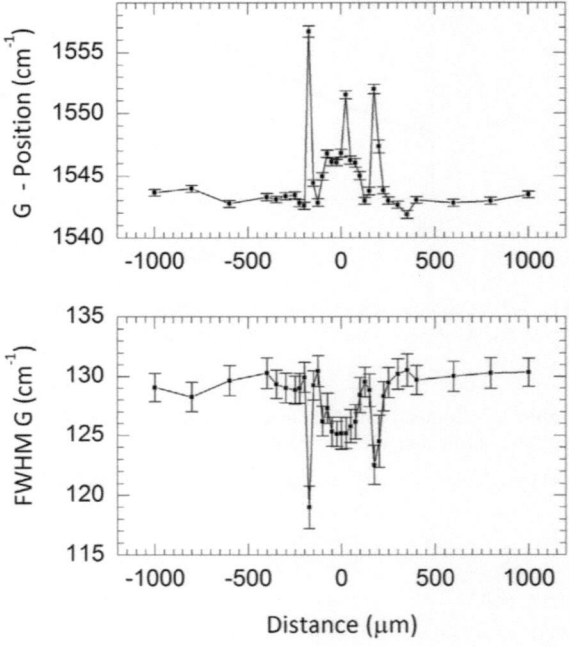

Fig. 7.8 Variation of the G-band position and full-width at half maximum across the wear track of silicon and oxygen incorporated DLC coating. Reprinted with permission from [1]

7.5 Fourier Transform Infrared Spectroscopy

Fourier transform infrared spectroscopy (FTIR) is an instrumental technique that facilitates the identification of the chemical substances and functional groups present in organic and inorganic compounds, irrespective of their solid, liquid or gaseous state. The atoms within chemical compounds are subject to constant movement and vibration, with each vibration occurring at a unique frequency that is specific to the chemical bond and compound. These frequencies correspond to those of light within the infrared region of the electromagnetic spectrum, typically spanning from 4000 to 600 cm^{-1}. The measurement principle is predicated on the irradiation of the sample with IR radiation across a broad spectral range, thereby inducing interaction between the IR light and the sample. The selective absorption of IR radiation results in the modulation of molecular vibrations, leading to alterations in the vibrational energy of the molecules. As this energy is quantised, only radiation with specific energies, characteristic of the functional groups performing vibrations, is absorbed. It is possible to analyse both the light reflected from the sample and the light that passes through the sample after interacting with it. The resulting spectrum represents the molecular absorption and transmission, creating a molecular fingerprint of the sample, as no two unique molecular structures produce the same infrared spectrum [16]. FTIR technique is frequently considered a complementary method to XPS. While both are regarded as powerful analytical techniques, they are suited to different types of analysis. FTIR provides information about molecular structure and functional groups, while XPS is primarily used for surface analysis, providing details about elemental composition and chemical states at the surface of a material [17, 18].

In the analysis of the surface chemical structure of carbon coatings for tribological applications, the most common technique utilised is Attenuated Total Reflectance FTIR (ATR-FTIR) spectroscopy, given that the analysed coatings are typically deposited on metal alloy samples. This technique involves the transmission of light through a crystal upon which the tested surface is pressed. The light undergoes total internal reflection, at least once, at the crystal-sample interface, after which it is directed towards the FTIR detector. FTIR studies of post-mortem analysis of wear tracks are especially useful for the analysis of functional groups formed on the coating surface as the result of chemical interactions between the coating material and the surrounding environment. These usually include the tribological tests in liquid lubricants [8, 17, 19] and simulated body fluids [20, 21]. FTIR analysis of wear tracks obtained after friction tests of silicon incorporated carbon coatings in simulated body fluid (SFB) and bovine serum albumin (BSA) confirmed, that their tribological parameters are governed by the presence of oxygen rather than the changing concentration of silicon [22]. The FTIR spectra of wear tracks of the coatings deposited using different silicon precursors (tetramethylsilane (TMS) and hexamethyldisiloxane (HMDSO)), tested in SBF and BSA environments, are presented in Fig. 7.9. The registered characteristic IR absorption bands are summarised in Table 7.2.

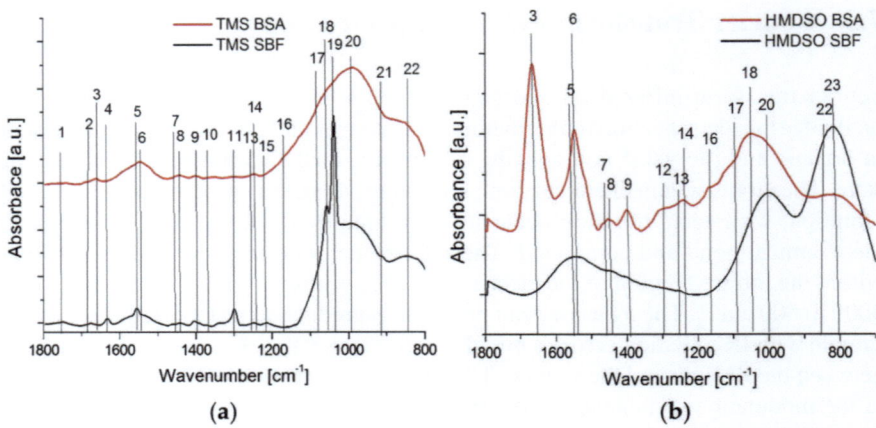

Fig. 7.9 FTIR spectra of silicon-incorporated DLC coatings **a** deposited using TMS and **b** deposited using HMDSO, after ball-on-disc tests in SBF and BSA environments. Reprinted with permission under Creative Commons CC-BY from [23]

A set of distinctive bands, indicative of albumin derived from N–H, C–N and C–C–N bonds, were identified in HMDSO coatings, suggesting high surface adhesion of proteins. In contrast, TMS coatings exhibited significantly lower band intensities. This observation underpins the distinct tribological behaviour exhibited by these two coating types. Specifically, HMDSO coatings exhibited higher values of the coefficient of friction in the BSA environment, accompanied by a significantly lower wear rate, in contrast to TMS coatings, which exhibited a low coefficient of friction and high wear rate. The divalent nature of oxygen in HMDSO coatings was found to enhance their ability to bind protein molecules, thereby protecting the surface against excessive wear [22].

7.6 In-Situ Studies of Tribological Processes of Carbon Based Low Friction Coatings

The advancement of technology, particularly in the domain of solid-state physics, has resulted in a considerable reduction in the size of devices while concurrently enhancing their resolution and sampling frequency. This progression has yielded substantial opportunities for conducting in-situ studies of the dynamics of tribological processes, encompassing progressive wear or tribochemical transformations of carbon structures induced during friction contact. The configuration of the sample and the friction pair allows for the placement of additional measuring equipment without affecting the tribological process itself. A key benefit of these solutions is the elimination of the need for interruptions during testing, as each immobilisation

Table 7.2 Characteristic IR absorption bands in the range of 1800–700 cm^{-1} registered in the wear track of the analyzed silicon-incorporated DLC coatings [23]

No.	Wavenumber [cm^{-1}]	Vibrating mode [cm^{-1}]	TMS SBF	TMS BSA	HMDSO SBF	HMDSO BSA
1	1750	C = O (stretch)	+	−	−	−
2	1680	C = O (stretch)	+	+	−	−
3	1659	C = O, C–N (stretch) (amid I)	−	+	−	+ (strong)
4	1630	C = C (stretch)	+	−	−	−
5	1550	C = O (stretch)	+ (strong)	+	+	+
6	1540	N–H, C–N (deformation) (amid II)	−	+	−	+ (strong)
7	1460	CH$_3$ (deformation)	+	+	+	+
8	1440	C–OH/Si–OH (Bend)	+	+	+	+
9	1390	C–N (stretch) (amide III band)	−	+	−	+ (strong)
10	1370	SiOCOCH$_3$	+ (weak)	+ (weak)	−	−
11	1305	C = O (stretch)	+ (strong)	−	−	−
12	1299	C–N (stretch) (amide III)	−	−	−	+
13	1250	Si–CH$_3$ (bend)	+	+	+	+
14	1240	C–N (stretch)	−	+	−	+
15	1213	C–O–C (stretch)	+	−	−	−
16	1170	C–C–N (stretch)	−	+	−	+
17	1080	Si–O–C/ R1-Si–O–Si-R2	+	+	+	+
18	1057	(stretch)	+ (strong)	+	+	+
19	1038	Si–O/Si–CH$_2$–Si (stretch)	+ (strong)	+	−	−
20	1000	Si–O–Si (stretch)	+	+	+	+
21	920	Si–O(stretch)	+	−	−	−
22	857	CH = CH (deformation)	+	+	+	+
23	802	Si–O–C (stretch)	+	+	+	+

of the friction pair alters the conditions of the entire tribosystem. A particularly noteworthy illustration of this is the utilisation of the phenomenon of electromagnetic wave interference on the contact surface, facilitated by employing a counter-sample composed of a transparent material. This methodology empowers real-time identification of the chemical composition and thickness of the transition layer during tribological tests of low-friction carbon-based coatings under diverse conditions [24, 25]. The extension of this solution is a device that enables the regulation of the temperature of the friction contact, equipped with a high-temperature laser light supply and detection system, as illustrated in Fig. 7.10. The equipment facilitates the collection of Raman spectra during wear tests at temperatures ranging from 25 to 700 °C. This includes the following: (i) the identification of the chemical structure of wear tracks immediately before permanent damage of the analysed surface, (ii) the assessment of the effect of wear on the chemical composition of the surface, and (iii) the detection of newly formed phases resulting from tribochemical processes occurring at elevated temperatures [26].

Figure 7.11 illustrates a schematic diagram of the Multi-Probe Chromatic Aberration System (MPCAS) which facilitates the acquisition of three-dimensional maps of the wear track as the test progresses. The system utilises white light illumination from each probe, which is then imaged by a chromatic objective and analysed by a spectrometer. The distance of the chromatic objective from the sample surface at the illuminated point corresponds to the average wavelength of light detected by the spectrometer. The MPCAS system is capable of operating at a frequency of 2 kHz, thereby enabling the combination of multiple sequential profiles to obtain a real-time wear scar map without the need to interrupt the test [27].

An interesting example of utilisation of the transmission electron microscopy technique for in-situ studies of tribological transformations of sp^3 hybridized carbon

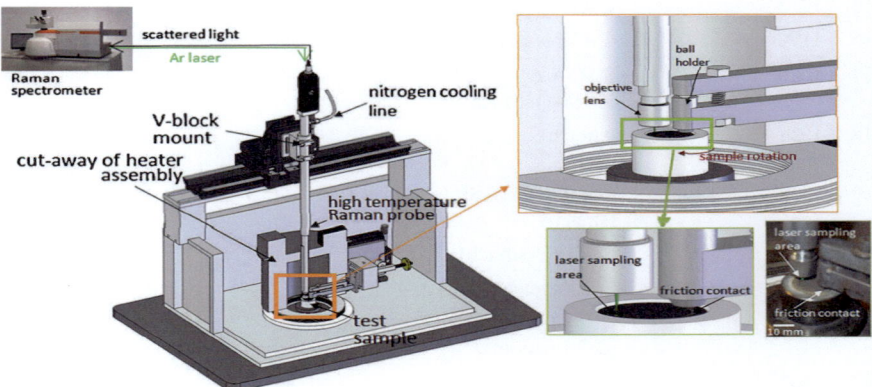

Fig. 7.10 Schematics and photograph of tribometer with the in situ Raman testing apparatus. Reprinted with permission from [26]

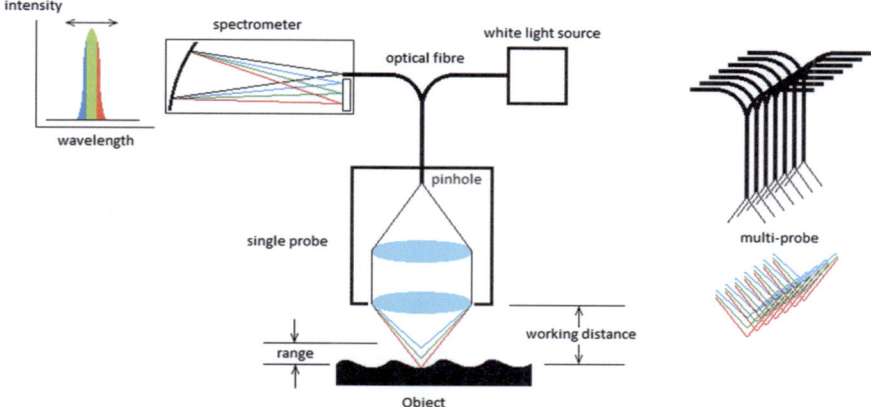

Fig. 7.11 Schematics of principle of operation of multi-probe chromatic aberration system. Reprinted with permission form [27]

structures was reported in [28]. In the presented configuration, the mobile sample holder facilitates mutual, relative movement between the friction system formed by the tungsten nanoprobe and the electron-transparent sample surface that has been modified with a carbon coating. Utilising electron energy loss spectroscopy (EELS) spectra acquired during individual sliding cycles, the researchers observed processes analogous to graphitization within the carbon coating.

Another solution is based on the use of a scanning probe microscope (SCA) in conjunction with Raman spectroscopy and a nanoindentation system. The device facilitates the execution of tribological tests of coatings at variable load values, whilst simultaneously conducting analyses of abrasion profiles and enabling continuous investigation of the chemical structure of the substrate material [29].

References

1. Jedrzejczak, A., Kolodziejczyk, L., Szymanski, W., Piwonski, I., Cichomski, M., Kisielewska, A., Dudek, M., Batory, D.: Friction and wear of a-C:H:SiOx coatings in combination with AISI 316L and ZrO_2 counterbodies. Tribol. Int. **112**, 155–162 (2017). https://doi.org/10.1016/j.triboint.2017.03.026
2. Tseng, A.A.: Recent developments in nanofabrication using focused ion beams. Small **1**, 924–939 (2005). https://doi.org/10.1002/smll.200500113
3. Höflich, K., Hobler, G., Allen, F.I., Wirtz, T., Rius, G., McElwee-White, L., Krasheninnikov, A.V., Schmidt, M., Utke, I., Klingner, N., Osenberg, M., Córdoba, R., Djurabekova, F., Manke, I., Moll, P., Manoccio, M., De Teresa, J.M., Bischoff, L., Michler, J., De Castro, O., Delobbe, A., Dunne, P., Dobrovolskiy, O.V., Frese, N., Gölzhäuser, A., Mazarov, P., Koelle, D., Möller, W., Pérez-Murano, F., Philipp, P., Vollnhals, F., Hlawacek, G.: Roadmap for focused ion beam technologies. Appl. Phys. Rev. **10** (2023). https://doi.org/10.1063/5.0162597

4. Bird, A., Yang, L., Wu, G., Inkson, B.J.: Failure mechanisms of diamond like carbon coatings characterised by in situ SEM scratch testing. Wear 530–531 (2023). https://doi.org/10.1016/j.wear.2023.205034
5. Stevie, F.A., Donley, C.L.: Introduction to x-ray photoelectron spectroscopy. J. Vacuum Sci. Technol. A Vacuum Surf. Films **38** (2020). https://doi.org/10.1116/6.0000412
6. Pinder, J.W., Major, G.H., Baer, D.R., Terry, J., Whitten, J.E., Čechal, J., Crossman, J.D., Lizarbe, A.J., Jafari, S., Easton, C.D., Baltrusaitis, J., van Spronsen, M.A., Linford, M.R.: Avoiding common errors in X-ray photoelectron spectroscopy data collection and analysis, and properly reporting instrument parameters. Appl. Surf. Sci. Adv. **19**, 100534 (2024). https://doi.org/10.1016/j.apsadv.2023.100534
7. Omiya, T., Cavaleiro, A., Figueiredo, N., Gouttebaron, R., Felten, A., Ferreira, F.: Sustainable lubrication through Gd DLC films and ionic liquids for wear and corrosion resistance. Tribol. Int. **200** (2024). https://doi.org/10.1016/j.triboint.2024.110130
8. Guan, Y., Marquis, E., Righi, M.C., Galipaud, J., Dubreuil, F., Dufils, J., Macron, E., Dassenoy, F., de Barros Bouchet, M.I.: Friction control by load-induced structure modification of over-based detergent in fully formulated lubricant. Tribol. Int. **192** (2024). https://doi.org/10.1016/j.triboint.2024.109307
9. Zhang, R., Lee, W.Y., Umehara, N., Tokoroyama, T., Murashima, M., Takimoto, Y.: The development of B/Cr co-doped DLC coating by FCVA deposition system and its tribological properties at 300 °C. Surf. Coat. Technol. **487** (2024). https://doi.org/10.1016/j.surfcoat.2024.130968
10. Batory, D., Jedrzejczak, A., Kaczorowski, W., Szymanski, W., Kolodziejczyk, L., Clapa, M., Niedzielski, P.: Influence of the process parameters on the characteristics of silicon-incorporated a-C:H:SiOx coatings. Surf. Coat. Technol. **271**, 112–118 (2015). https://doi.org/10.1016/j.surfcoat.2014.12.073
11. Casiraghi, C., Piazza, F., Ferrari, A.C., Grambole, D., Robertson, J.: Bonding in hydrogenated diamond-like carbon by Raman spectroscopy. In: Diamond and Related Materials, pp. 1098–1102 (2005)
12. Ferrari, A.C., Robertson, J.: Interpretation of Raman spectra of disordered and amorphous carbon. Phys. Rev. B **61**, 14095–14107 (2000). https://doi.org/10.1103/PhysRevB.61.14095
13. Robertson, J.: Diamond-like amorphous carbon. Mater. Sci. Eng. R. Rep. **37**, 129–281 (2002). https://doi.org/10.1016/S0927-796X(02)00005-0
14. Batory, D., Jedrzejczak, A., Kaczorowski, W., Kolodziejczyk, L., Burnat, B.: The effect of Si incorporation on the corrosion resistance of a-C:H:SiOx coatings. Diam. Relat. Mater. **67** (2016). https://doi.org/10.1016/j.diamond.2015.12.002
15. Endrino, J.L., Escobar Galindo, R., Zhang, H.-S., Allen, M., Gago, R., Espinosa, A., Anders, A.: Structure and properties of silver-containing a-C(H) films deposited by plasma immersion ion implantation. Surf. Coat. Technol. **202**, 3675–3682 (2008). https://doi.org/10.1016/j.surfcoat.2008.01.011
16. Smith, B.C.: Fundamentals of Fourier Transform Infrared Spectroscopy. CRC Press (2011)
17. Čoga, L., Akbari, S., Kovač, J., Kalin, M.: Differences in nano-topography and tribochemistry of ZDDP tribofilms from variations in contact configuration with steel and DLC surfaces. Friction **10**, 296–315 (2022). https://doi.org/10.1007/s40544-021-0491-7
18. Sharifahmadian, O., Mahboubi, F.: A comparative study of microstructural and tribological properties of N-DLC/DLC double layer and single layer coatings deposited by DC-pulsed PACVD process. Ceram. Int. **45**, 7736–7742 (2019). https://doi.org/10.1016/j.ceramint.2019.01.076
19. Zhang, T.F., Xie, D., Huang, N., Leng, Y.: The effect of hydrogen on the tribological behavior of diamond like carbon (DLC) coatings sliding against Al_2O_3 in water environment. Surf. Coat. Technol. **320**, 619–623 (2017). https://doi.org/10.1016/j.surfcoat.2016.10.045
20. Jing, P.P., Su, Y.H., Li, Y.X., Liang, W.L., Leng, Y.X.: Mechanism of protein biofilm formation on Ag-DLC films prepared for application in joint implants. Surf. Coat. Technol. **422** (2021). https://doi.org/10.1016/j.surfcoat.2021.127553

References

21. Zhang, T.F., Liu, B., Wu, B.J., Liu, J., Sun, H., Leng, Y.X., Huang, N.: The stability of DLC film on nitrided CoCrMo alloy in phosphate buffer solution. Appl. Surf. Sci. **308**, 100–105 (2014). https://doi.org/10.1016/j.apsusc.2014.04.117
22. Jedrzejczak, A., Szymanski, W., Kolodziejczyk, L., Sobczyk-Guzenda, A., Kaczorowski, W., Grabarczyk, J., Niedzielski, P., Kolodziejczyk, A., Batory, D.: Tribological characteristics of a-C:H: Si and a-C:H:SiOx coatings tested in simulated body fluid and protein environment. Materials **15**, 2082 (2022). https://doi.org/10.3390/ma15062082
23. Jedrzejczak, A., Szymanski, W., Kolodziejczyk, L., Sobczyk-Guzenda, A., Kaczorowski, W., Grabarczyk, J., Niedzielski, P., Kolodziejczyk, A., Batory, D.: Tribological characteristics of a-C:H:Si and a-C:H:SiOx coatings tested in simulated body fluid and protein environment. Materials **15** (2022). https://doi.org/10.3390/ma15062082
24. Scharf, T.W., Singer, I.L.: Quantification of the thickness of carbon transfer films using raman tribometry. Tribol. Lett. **14**, 137–145 (2003). https://doi.org/10.1023/A:1021942822261
25. Chromik, R.R., Baker, C.C., Voevodin, A.A., Wahl, K.J.: In situ tribometry of solid lubricant nanocomposite coatings. Wear **262**, 1239–1252 (2007). https://doi.org/10.1016/j.wear.2007.01.001
26. Muratore, C., Bultman, J.E., Aouadi, S.M., Voevodin, A.A.: In situ Raman spectroscopy for examination of high temperature tribological processes. Wear **270**, 140–145 (2011). https://doi.org/10.1016/j.wear.2010.07.012
27. Gee, M., Kamps, T., Woolliams, P., Nunn, J., Mingard, K.: In situ real time observation of tribological behaviour of coatings. Surf. Coat. Technol. **442** (2022). https://doi.org/10.1016/j.surfcoat.2022.128233
28. Merkle, A.P., Erdemir, A., Eryilmaz, O.L., Johnson, J.A., Marks, L.D: In situ TEM studies of tribo-induced bonding modifications in near-frictionless carbon films. Carbon N Y **48**, 587–591 (2010). https://doi.org/10.1016/j.carbon.2009.08.036
29. Rouhani, M., Hobley, J., Hong, F.C.N., Jeng, Y.R.: Novel spatially coordinated in-situ Raman and nanoscale wear analysis of FCVA-deposited DLC film. AIP Adv. **9** (2019). https://doi.org/10.1063/1.5107474

Chapter 8
Tribocorrosion and Biotribocorrosion of Carbon Based Low Friction Coatings

Abstract The chapter expounds upon tribo and biotribocorrosion phenomena. The principle of the test is elucidated, and an example diagram of a tribometer equipped with an electrochemical cell is presented. The methodology of measuring and analysing the obtained results in relation to total wear is explained. Finally, a number of solutions aimed at improving wear lifetime and electrochemical characteristics of DLC layers are presented.

Tribocorrosion is defined as the surface transformations resulting from the interaction of mechanical loading and chemical/electrochemical reactions that occur between various elements of a tribosystem exposed to a corrosive environment. It is a simultaneous mechanism involving the mechanical and chemical interactions of body, counterbody, interfacial medium and the environment, including friction, lubrication, wear and tribologically activated chemical and electrochemical reactions. As a result, the behaviour of tribocorrosion cannot be reliably deduced from the outcomes of individual wear and corrosion experiments alone [1, 2]. Tribocorrosion, defined as the corrosion of a material caused by friction and movement, has the potential to occur in a wide range of scenarios involving any form of friction contact in an aggressive or corrosive environment. Illustrative examples include steel conveyors operating in high relative humidity, components of offshore infrastructure exposed to high humidity or high salinity environments, power generation components operating under high pressure and high temperature conditions, erosion-corrosion of slurry pipes, and food processing applications. Recent years have seen a marked increase in the dynamic growth of the medical implant market. This has been accompanied by a surge in research activity in the field of tribocorrosion in biological environments (Biotribocorrosion). A significant proportion of this research has focused on the design of medical implants for load-bearing applications, such as knee and hip joint prostheses [3]. Biotribocorrosion processes encompass a range of phenomena, including oxidation, degradation of the passive layer, and the generation and transfer

Fig. 8.1 Schematic representation of the configuration of reciprocating tribometer in conjunction with an electrochemical cell. Reprinted with permission from [14]

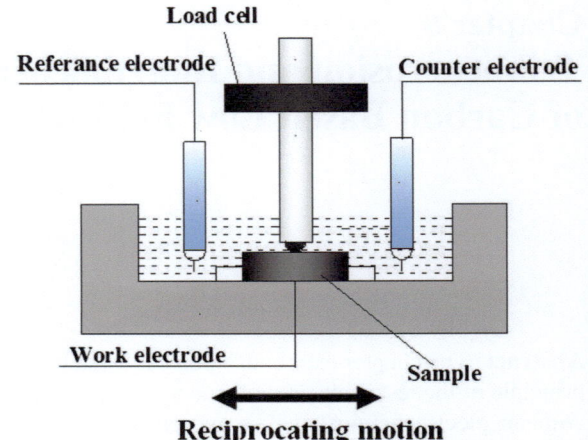

of corrosion products. The corrosion-related mechanical wear contributes significantly to the formation and migration of metal debris and ions. These ions are subsequently absorbed and accumulate in the surrounding tissues [4, 5]. It is estimated that between 10 and 30% of all endoprosthesis revisions are due to aseptic loosening, mainly caused by wear products formed in the contact couple [6]. Therefore, in the case of tribological systems operating in aggressive environmental conditions, it is important to understand material degradation processes to achieve longer service life and better safety issues.

DLC coatings are among the most extensively studied thin film materials employed to enhance implant performance, particularly due to their well-documented high corrosion resistance [7, 8], excellent biocompatibility [9–11], low wear and low friction coefficient (between 0.05 and 0.2) while sliding against most materials [12]. However, it is only through comprehensive characterisation of the mechanical and chemical interactions between the coating and the surrounding environment, as well as the counterbody material under friction and wear conditions, that the overall performance of the coating can be definitively ascertained and its potential applications can be fully realised.

An illustrative representation of a system that facilitates the execution of tribocorrosion tests on DLC-modified metallic substrates is provided in Fig. 8.1. In the proposed methodology, the DLC-modified substrate functions as the working electrode. In addition, reference and counter electrodes are immersed in the corrosive solution. The setup comprises a specially designed sample holder and load system, which facilitates the filling of the corrosion cell with a corrosive environment, including simulated body fluids [13–17], artificial sea water [18], simple NaCl solutions [19, 20] and other.

A chemically inert, non-conductive, ceramic-based material is typically employed as the counterbody. In addition to the assessment of the friction coefficient, a range of analytical techniques, including Open Circuit Potential (OCP), Electrochemical

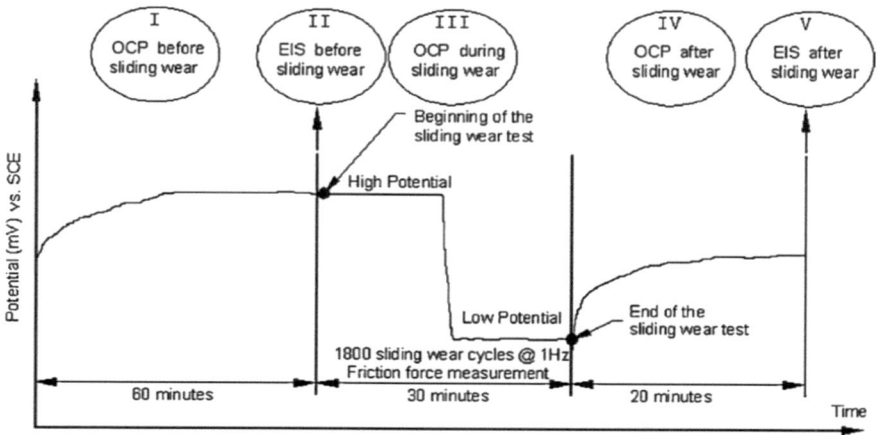

Fig. 8.2 Sequence of operations during the tribocorrosion test illustrated by the OCP evolution. Reprinted with permission form [19]

Impedance Spectroscopy (EIS) and Potentiodynamic Polarization Tests (PPT), may be utilised. These tests are performed using a potentiostat–galvanostat, which is integrated with a computer-controlled data acquisition system. Azzi et al. [19] proposed a sequence of operations during the tribcorrosion testing of DLC modified 316L austenitic steel (see Fig. 8.2).

The results of the test in terms of electrochemical testing are most often expressed in the form of EIS and OCP values. Changes in EIS can be interpreted by means of a simple model based on an equivalent electrical circuit of the impedance of the specimen before and after the tribocorrosion experiment. By contrast, the OCP is constantly measured during the entire test and thus provides a general overview of the changes in the behaviour of the tested material under the influence of friction and wear processes in an aggressive environment. A substantial decrease in the OCP value is generally observed when the passive layer is removed from the wear track, thereby exposing fresh material to the corrosive solution [21].

In the context of standard tribological tests, the wear rate is determined postmortem through the measurement of the wear track profile. In the case of tribocorrosion tests, however, the wear loss volume consists of three components. The total wear loss volume (W_{total}) under tribo-corrosion conditions is the sum of mechanical volume losses (W_{mech}), chemical volume losses (W_{chem}) and synergistic volume losses (W_{syn}), according to the following equation [18, 22].

$$W_{total} = W_{mech} + W_{chem} + W_{syn} \tag{8.1}$$

Cathodic polarization of the surface has been shown to suppress corrosion reactions. Consequently, material loss during wear testing is attributable only to mechanical wear W_{mech}. Both W_{total} and W_{mech} wear rates can be easily determined by the analysis of the volume of wear track after the test.

The chemical wear volume W_{chem} is determined by the Faraday's law:

$$W_{chem} = \frac{ItM}{nF\rho} \tag{8.2}$$

where:

I   corrosion current obtained from the polarization curve without sliding [A]
t   time of sliding contact [s]
F   Faraday's constant
M   atomic mass of the sample
n   is the charge number for the oxidation reaction
ρ   is the density of the samples

With regard to the final component, M_{syn}, this is defined as the synergistic effect of wear and corrosion. Specifically, friction and wear modify the sensitivity of materials to corrosion, and conversely, corrosion may alter the conditions of wear. Therefore, M_{syn} can be expressed as the sum of two components: the corrosion loss due to tribological action, otherwise known as wear-enhanced corrosion, and the wear loss due to corrosion, otherwise known as corrosion-enhanced wear [23]. In the majority of cases, these two components are considered as a single entity (M_{syn}), which is subsequently calculated as the total mechanical and chemical wear components can be analytically determined. However, this approach does not permit the determination of the contribution of each component to the total degradation of the material. Consequently, the material loss due to wear-enhanced corrosion can also be calculated. To this end, the average current during wear is first calculated. The resulting difference between the average current during wear and the current measured before wear is then used to calculate the wear-enhanced corrosion current. This is followed by the application of Faraday's law once more. The corrosion-enhanced wear is then the difference between the total material loss and all the other losses [22].

The well-documented history of failure of DLC-coated femoral heads implanted between 1993 and 1995 was revealed approximately 8.5 years after the implantation. Approximately 50% of the hip joints required revision due to aseptic loosening. A thorough examination of the failed femoral heads revealed that coating imperfections were the primary cause of failure. The presence of small-diameter pinholes can be attributed to dust, substrate flaws, and particles generated during the deposition process. Subsequent infiltration of these pinholes with body fluids led to extensive corrosion of the adhesion-promoting Si interlayer at the interface between the coating and the substrate. This ultimately resulted in delamination of the coating and severe wear of the polyethylene cups [24]. Carbon coatings are also known for their high residual stress and poor adhesion to metallic materials, particularly when there are significant differences in the mechanical properties of the coating and the substrate

[25–27]. Stress and strain-induced defects, such as cracks and local delamination, can also act as easy infiltration routes for corrosive environments. Consequently, the wear life of DLC coatings is significantly diminished in tribocorrosion tests in comparison to experiments conducted under dry friction conditions. Notwithstanding this, the persistent interest of industry and academia in the exceptional properties of carbon coatings ensures their ongoing investigation in numerous laboratories and R&D centres globally.

A number of solutions aimed at improving wear lifetime and electrochemical characteristics of DLC layers are reported. Metallic elements have commonly been incorporated into amorphous carbon matrix to decrease residual stress, improve adhesion and other mechanical and tribological properties. It has been demonstrated that doping processes may also reduce the occurrence of pinholes on the surface. Tungsten-incorporated carbon coatings exhibited significantly increased corrosion and tribocorrosion resistance in artificial seawater. It was observed that coatings with 0.9 at.% W exhibited a significantly reduced number of pinholes on the surface while maintaining the highest hardness value, highlighting the potential of tungsten-doped carbon coatings in enhancing corrosion and tribocorrosion resistance [18]. DLC-Ag coatings exhibited a significantly lower and more stable friction coefficient, in addition to providing enhanced surface protection against tribocorrosion in comparison to DLC layers. The incorporation of silver nanoparticles into the DLC matrix facilitated the synthesis of a compact structure with reduced residual stress, thereby preventing the infiltration and corrosion of salt ions [16].

Multilayer structures have been shown to be capable of neutralising defects, such as pores, pinholes and crevices, which occur in single-layer DLC films. This results in the prolongation or obstruction of the ion transportation path. The composition of the multilayer film is a crucial factor in this regard. Silicon incorporated multilayer DLC coatings have been demonstrated to exhibit superior barrier properties. In addition to the multilayer structure, the formation of Si oxides has been observed, which provide a more effective diffusion barrier for corrosive solutions [28]. The tribocorrosion resistance of DLC coatings under low load conditions is primarily contingent on the characteristics of the top DLC layer. However, under conditions of exposure to elevated loads, a marked alteration in the coating's behaviour becomes evident. In such cases, the mechanical properties of the substrate material and the interface adhesion strength between the top layer and the substrate become of particular significance. It is imperative to note that in the absence of adequate support for the hard and brittle carbon coating, irrespective of its superior corrosion and tribological properties, the system is vulnerable to failure under high loads. This phenomenon can be attributed to the substantial elastic deformation of the substrate material under the applied load, which subsequently leads to the brittle fracture of the top coating. This, in turn, facilitates unimpeded access to the base material by the aggressive environment [20].

One potential solution is the deposition of Cr/CrC_x adhesion-promoting interlayers, which have been shown to effectively mitigate the formation of cracks under conditions of high stress, corrosion and wear. In Fig. 8.3, a schematic diagram of tribocorrosion mechanisms for different variations of the interlayer structure and

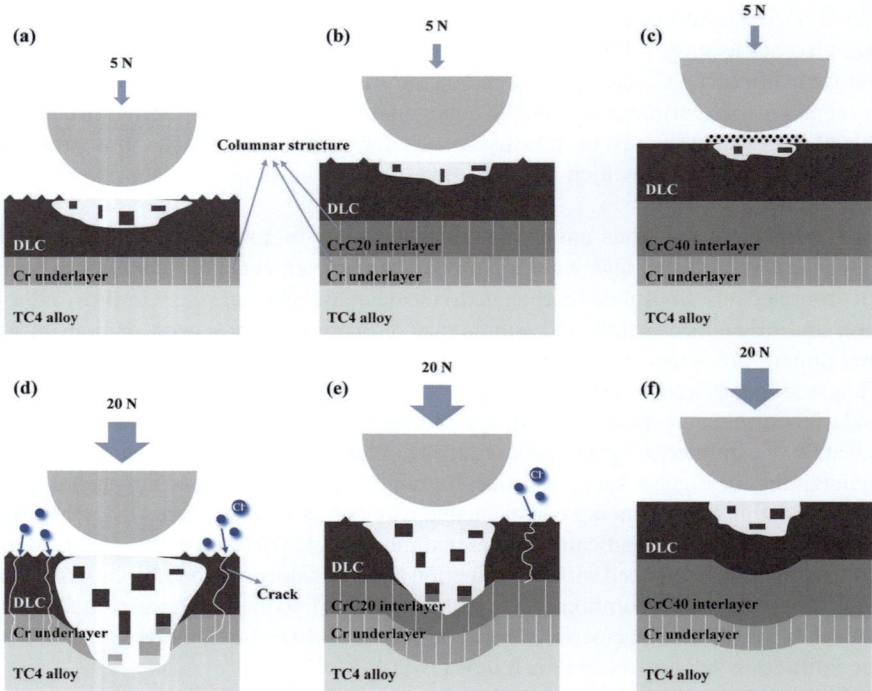

Fig. 8.3 Schematic diagram of tribocorrosion mechanism for different DLC coatings under 5 and 20 N load conditions. Reprinted with permission form [29]

material under load of 5 and 20 N is presented. It is evident that the optimized interlayer composition and thickness provide a sufficient support for the DLC coating. The absence of cracks in the top coating serves to act as a barrier against the permeation of corrosive media, thereby protecting the substrate [29].

Grabarczyk et al. investigated the influence of thermo-chemical treatment of Ti–6A–l4V alloy on the tribocorrosion properties of DLC coatings deposited on the top and tested in PBS solution. Their research revealed that a two-stage modification process, involving glow discharge oxidation of Ti–6A–l4V substrate and the subsequent deposition of a carbon coating, resulted in enhanced tribocorrosion behaviour. This included a reduction in the coefficient of friction, an increase in hardness, and a wear rate that was barely measurable. This emphasises the importance of ensuring that the properties of the coating and substrate are adequately matched, especially in the context of highly loaded, tribocorrosion-related applications [30].

Finally, the tribocorrosion tests of the DLC-modified CoCrMo alloy revealed, that the absorbed protein layer on the DLC surface plays a positive role in reducing wear and metal ion release. During the polarization tests under sliding conditions a higher breakdown potential and a lower passive current density were registered for DLC coatings tested in bovine calf serum compared to those in 0.9% NaCl solution [14].

References

1. Mischler, S., Spiegel, A., Stemp, M., Landolt, D.: Influence of passivity on the tribocorrosion of carbon steel in aqueous solutions (2001)
2. Landolt, D., Mischler, S.: Tribocorrosion of Passive Metals and Coatings. Woodhead Publishing Limited (2011)
3. Merola, M., Affatato, S.: Materials for hip prostheses: a review of wear and loading considerations. Materials **12** (2019). https://doi.org/10.3390/ma12030495
4. Bauer, S., Schmuki, P., von der Mark, K., Park, J.: Engineering biocompatible implant surfaces: Part I: materials and surfaces. Prog. Mater. Sci. **58**, 261–326 (2013). https://doi.org/10.1016/j.pmatsci.2012.09.001
5. Hesketh, J., Meng, Q., Dowson, D., Neville, A.: Biotribocorrosion of metal-on-metal hip replacements: How surface degradation can influence metal ion formation. Tribol. Int. **65**, 128–137 (2013). https://doi.org/10.1016/j.triboint.2013.02.025
6. Kenney, C., Dick, S., Lea, J., Liu, J., Ebraheim, N.A.: A systematic review of the causes of failure of Revision Total Hip Arthroplasty. J. Orthop. **16**, 393–395 (2019). https://doi.org/10.1016/j.jor.2019.04.011
7. Batory, D., Blaszczyk, T., Clapa, M., Mitura, S.: Investigation of anti-corrosion properties of Ti:C gradient layers manufactured in hybrid deposition system. J. Mater. Sci. **43** (2008). https://doi.org/10.1007/s10853-007-2393-0
8. Batory, D., Jedrzejczak, A., Kaczorowski, W., Kolodziejczyk, L., Burnat, B.: The effect of Si incorporation on the corrosion resistance of a-C:H:SiOx coatings. Diam. Relat. Mater. **67** (2016). https://doi.org/10.1016/j.diamond.2015.12.002
9. Gotzmann, G., Beckmann, J., Scholz, B., Herrmann, U., Wetzel, C.: Low-energy electron-beam modification of DLC coatings reduces cell count while maintaining biocompatibility. Surf. Coat. Technol. **336**, 34–38 (2018). https://doi.org/10.1016/j.surfcoat.2017.09.024
10. Jamesh, M.I., Li, P., Bilek, M.M.M., Boxman, R.L., McKenzie, D.R., Chu, P.K.: Evaluation of corrosion resistance and cytocompatibility of graded metal carbon film on Ti and NiTi prepared by hybrid cathodic arc/glow discharge plasma-assisted chemical vapor deposition. Corros. Sci. **97**, 126–138 (2015). https://doi.org/10.1016/j.corsci.2015.04.022
11. Hang, R., Qi, Y.: Diamond and related materials a study of biotribological behavior of DLC coatings and its in fl uence to human serum albumin. Diam. Relat. Mater. **19**, 62–66 (2010). https://doi.org/10.1016/j.diamond.2009.11.008
12. Hauert, R., Thorwarth, K., Thorwarth, G.: An overview on diamond-like carbon coatings in medical applications. Surf. Coat. Technol. **233**, 119–130 (2013). https://doi.org/10.1016/j.surfcoat.2013.04.015
13. Arslan, E., Totik, Y., Efeoglu, I.: The investigation of the tribocorrosion properties of DLC coatings deposited on Ti–6A–l4V alloys by CFUBMS. Prog. Org. Coat. **74**, 768–771 (2012). https://doi.org/10.1016/j.porgcoat.2011.10.023
14. Liu, J., Wang, X., Wu, B.J., Zhang, T.F., Leng, Y.X., Huang, N.: Tribocorrosion behavior of DLC-coated CoCrMo alloy in simulated biological environment. Vacuum **92**, 39–43 (2013). https://doi.org/10.1016/j.vacuum.2012.11.017
15. Li, Y., Zhou, Z., He, Y.: Tribocorrosion and surface protection technology of titanium alloys: a review. Materials **17**, 65 (2023). https://doi.org/10.3390/ma17010065
16. Radi, P.A., Vieira, L., Leite, P., Trava-Airoldi, V.J., Massi, M., Reis, D.A.P.: Tribocorrosion studies on DLC films with silver nanoparticles for prosthesis applications. Surf Topogr. **12**, 015019 (2024). https://doi.org/10.1088/2051-672X/ad2ebe
17. Hinüber, C., Kleemann, C., Friederichs, R.J., Haubold, L., Scheibe, H.J., Schuelke, T., Boehlert, C., Baumann, M.J.: Biocompatibility and mechanical properties of diamond-like coatings on cobalt-chromium-molybdenum steel and titanium-aluminum-vanadium biomedical alloys. J. Biomed. Mater. Res. A **95A**, 388–400 (2010). https://doi.org/10.1002/jbm.a.32851
18. Cao, L., Liu, J., Wan, Y., Pu, J.: Corrosion and tribocorrosion behavior of W doped DLC coating in artificial seawater. Diam. Relat. Mater. **109** (2020). https://doi.org/10.1016/j.diamond.2020.108019

19. Azzi, M., Paquette, M., Szpunar, J.A., Klemberg-Sapieha, J.E., Martinu, L.: Tribocorrosion behaviour of DLC-coated 316L stainless steel. Wear **267**, 860–866 (2009). https://doi.org/10.1016/j.wear.2009.02.006
20. Zhang, Y., Li, H., Cui, L., Yang, W., Ma, G., Chen, R., Guo, P., Ke, P., Wang, A.: Comparative study on tribocorrosion behavior of hydrogenated/hydrogen-free amorphous carbon coated WC-based cermet in 3.5 wt% NaCl solution. Corros. Sci. **227**, 111738 (2024). https://doi.org/10.1016/j.corsci.2023.111738
21. Ponthiaux, P., Wenger, F., Drees, D., Celis, J.P.: Electrochemical techniques for studying tribocorrosion processes. Wear **256**, 459–468 (2004). https://doi.org/10.1016/S0043-1648(03)00556-8
22. Martin, É., Azzi, M., Salishchev, G.A., Szpunar, J.: Influence of microstructure and texture on the corrosion and tribocorrosion behavior of Ti–6Al–4V. Tribol. Int. **43**, 918–924 (2010). https://doi.org/10.1016/j.triboint.2009.12.055
23. Mischler, S.: Triboelectrochemical techniques and interpretation methods in tribocorrosion: a comparative evaluation. Tribol. Int. **41**, 573–583 (2008). https://doi.org/10.1016/j.triboint.2007.11.003
24. Hauert, R., Falub, C.V., Thorwarth, G., Thorwarth, K., Affolter, C., Stiefel, M., Podleska, L.E., Taeger, G.: Retrospective lifetime estimation of failed and explanted diamond-like carbon coated hip joint balls. Acta Biomater. **8**, 3170–3176 (2012). https://doi.org/10.1016/j.actbio.2012.04.016
25. Batory, D., Jedrzejczak, A., Szymanski, W., Niedzielski, P., Fijalkowski, M., Louda, P., Kotela, I., Hromadka, M., Musil, J.: Mechanical characterization of a-C:H:SiOx coatings synthesized using radio-frequency plasma-assisted chemical vapor deposition method. Thin Solid Films **590** (2015). https://doi.org/10.1016/j.tsf.2015.08.017
26. Batory, D., Szymanski, W., Clapa, M.: Mechanical and tribological properties of gradient a-C:H/Ti coatings. Mater. Sci. Poland **31**, 415–423 (2013). https://doi.org/10.2478/s13536-013-0121-9
27. Jedrzejczak, A., Kolodziejczyk, L., Szymanski, W., Piwonski, I., Cichomski, M., Kisielewska, A., Dudek, M., Batory, D.: Friction and wear of a-C:H:SiOx coatings in combination with AISI 316L and ZrO_2 counterbodies. Tribol. Int. **112**, 155–162 (2017). https://doi.org/10.1016/j.triboint.2017.03.026
28. Cui, M., Pu, J., Liang, J., Wang, L., Zhang, G., Xue, Q.: Corrosion and tribocorrosion performance of multilayer diamond-like carbon film in NaCl solution. RSC Adv. **5**, 104829–104840 (2015). https://doi.org/10.1039/c5ra21207c
29. Li, Y., Qi, C., Guo, Z., Zhang, D., Sun, H., Yang, S., Wan, Y.: Effects of CrC interlayer on tribocorrosion properties of DLC-coated TC_4 alloy in a 0.9 wt% NaCl solution. Diam. Relat. Mater. **149** (2024). https://doi.org/10.1016/j.diamond.2024.111538
30. Grabarczyk, J., Gaj, J., Pazik, B., Kaczorowski, W., Januszewicz, B.: Tribocorrosion behavior of Ti–6A–14V alloy after thermo-chemical treatment and DLC deposition for biomedical applications. Tribol. Int. **153** (2021). https://doi.org/10.1016/j.triboint.2020.106560

Chapter 9
Synergistic Effect of Carbon Based Coatings with Self-assembled Monolayers

Abstract This chapter explores the synergy of appropriately modified carbon coatings and self-assembled monolayers (SAMs), with a focus on their general characteristics and the fundamental conditions for successful modification, leading to well-adherent SAM layers on Si-DLC and Ti-DLC coatings. The analysis of Me-DLC-SAMs compositions has revealed their potential to mitigate friction, adhesion and wear. Finally, the proposed applications of the obtained Me-DLC-SAMs nanocomposite coatings are presented.

Despite the excellent tribological properties of low-friction carbon-based coatings, further optimisation of their performance properties is still a priority, as these are often highly application-specific. Back in the early 2010s, DLC coatings were considered to be "inert" with a low surface energy. Consequently, they were thought not to react with the various oil additives and/or to attract polar groups from the additives and the oil, which are the conventional mechanisms for lubricating steels and other metals [1]. Accordingly, previous studies have concentrated on the chemical aspects of DLC lubrication, primarily on the reactivity with different additives, investigating possible reaction products and their effects on tribological performance. Contrary to the prevailing belief at the time that carbon coatings were non-reactive and chemically inert with respect to additives, recent years have seen the discovery of direct chemical evidence that DLC coatings also react with conventional steel-tailored additives. It has been observed, however, that such interactions are weaker, and fewer reaction products have been formed (or identified in post-experiment analysis). Furthermore, the adsorption of tribochemical films onto the surface was much lower [2]. Additionally, as presented in Chap. 5, during this period, the modification of the properties of carbon coatings through the doping process, involving the introduction of various elements, had already become a prevalent technique for enhancing their overall performance in a comprehensive manner. Consequently, at the point at which the limit of possibilities for modifying the tribological properties of low-friction carbon-based coatings through further changes to their synthesis technology had been reached, alternative solutions were sought. In this regard, the use of technologies for doping carbon coatings with metals was identified as a promising solution.

Fig. 9.1 Representation of a SAM structure

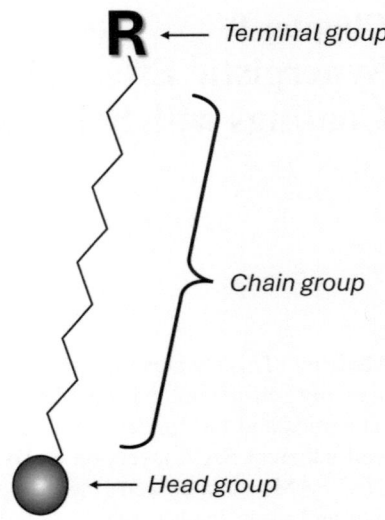

The advent of a plethora of technological solutions has engendered the capacity to regulate the concentration of dopant, in addition to the phase and chemical structure of the obtained coating. The resultant layers exhibited a metal-like character, contingent on the concentration of the doping element. The subsequent enhancement of their tribological properties was hypothesised to be achieved through the modification of the material with self-assembled monolayers (SAMs). They are a class of two-dimensional nanomaterials that are formed through the spontaneous assembly of molecular constituents onto the surface of various solids. This process results in the formation of highly ordered structures, grown by the adsorption of a one-molecule-thick layer onto the surface [3].

The structural composition of SAMs is divided into three distinct groups (see Fig. 9.1): a head group that reacts with a substrate, a backbone molecular chain group that connects the head and tail groups, and a terminal group that interacts with the outer surface of the film. This structural arrangement facilitates their utilisation as thin layers deposited on various substrates. The optimal choice for each group will yield the monolayer exhibiting the optimal performance characteristics.

The utilisation of organosilanes and alkylphosphonic acids as lubricants is a subject of considerable interest and technological attractiveness, primarily due to the low adhesion, friction, and wear of self-assembled monolayers (SAMs) obtained on their basis, particularly under a millinewton load range. The organosilanes are known to adopt a structural configuration of $R_n SiX_{4-n}$, while alkylphosphonic acids adopt a $RPO(OH)_2$ formula. In these formulations, R represents an organic functional group and X is a hydrolyzable leaving group, such as $-Cl$, $-OCH_3$ or $-OH$. The stability of SAMs on different substrates such as native oxide surfaces of various metals, including magnesium, aluminium, copper, nickel, titanium, and their alloys, is determined by covalent bonds between molecules and the surface, and cross-links

Table 9.1 Compounds used in the deposition of self-assembled monolayers (last two rows indicate compounds used in the modification of Si-DLC, SiO$_x$ and TiO$_x$-DLC types of coatings, whereas the first two were used in modification of Si-DLC and SiO$_x$-DLC layers)

Compound name	Empirical formula	References
1H, 1H, 2H,2H—perfluorodecyltrichlorosilane (FDTS)	C$_{10}$H$_4$Cl$_3$F$_{17}$Si	[13, 14]
3, 3, 3-trifluoropropyltrichlorosilane (FPTS)	C$_3$H$_4$Cl$_3$F$_3$Si	[13, 14]
n-decylphosphonic acid (DP)	C$_{10}$H$_{23}$PO$_3$	[15–17]
1H,1H,2H,2H—perfluorodecylphosphonic acid (PFDP)	C$_{10}$H$_6$F$_{17}$O$_3$P	[15–17]

between neighboring molecules consisting of three or two reactive atoms in the head group. The interactions between the backbone chain groups of different molecules, such as van der Waals forces or hydrogen bonds, contribute to the formation and stability of the deposited layers, which can be hydrophilic or hydrophobic [4, 5].

When selecting a suitable substrate for subsequent modification with SAMs, it was essential to consider carbon-based coatings, which have a wide range of performance characteristics. The structure and chemical composition of the coatings also played a key role in facilitating the formation of robust and resilient chemical bonds with the molecules of the self-assembled monolayers. The investigation focused on DLC coatings doped with Si or SiO$_x$, primarily due to their exceptional performance characteristics, including low residual stress levels, high corrosion resistance, minimal friction coefficient, and high wear resistance [6–10]. Conversely, the chemical structure was the primary factor in selecting TiO$_x$-DLC [11, 12]. This choice was further substantiated by the documented stability of P–O–M (M = Al, Ti, Zr, Si) bonds formed between the substrate and alkylphosphonic acid-based SAMs, as well as the fact that covalent bonds are formed between the fluoroalkysilans and the substrate via Si–O–Si bridges. Compounds used for investigations possessed methyl or trifluoromethyl terminal groups (see Table 9.1).

In Fig. 9.2 is presented FTIR spectra of silicon and oxygen incorporated carbon coating subsequently modified with two types of fluoroalkysilanes.

It was found, that layers of fluoroalkylsilanes are present on the surface of Si-DLC coatings and are covalently bonded to the substrate by Si–O–Si bridges. Schematic interpretation of the formation of self-assembled monolayer of 3, 3, 3-trifluoropropyltrichlorosilane on the surface of silicon incorporated carbon coating is presented in Fig. 9.3.

A comparison of the efficiency of modification for FPTS and FDTS manufactured on three different substrates is presented in Fig. 9.4.

The silicon incorporated DLC has been shown to be more attractive than silicon wafer and pure DLC coating. This is due to the presence of dispersed native silicon forms (SiO$_x$) within the coating which act as active centres for chemical bonds, thereby facilitating enhanced chemical modification. There was also a clear difference in modification efficiency between FDTS and FPTS. According to the authors, a higher concentration of fluorine and a tendency for vertical polymerisation of FDTS

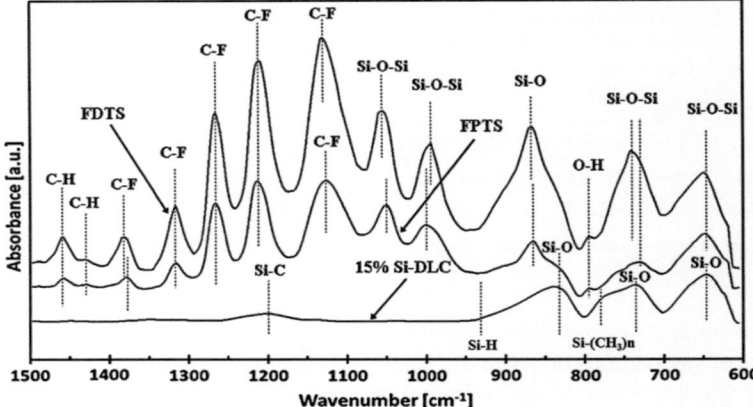

Fig. 9.2 FTIR spectra of Si-DLC(15% Si) coating before and after modification by FPTS and FDTS. Reprinted with permission from [13]

Fig. 9.3 The schematic representation of the formation of self-assembled monolayer on the surface of silicon incorporated carbon coating (the white dots symbolize Si atoms, which are the active centres for anchoring molecules of the modifier)

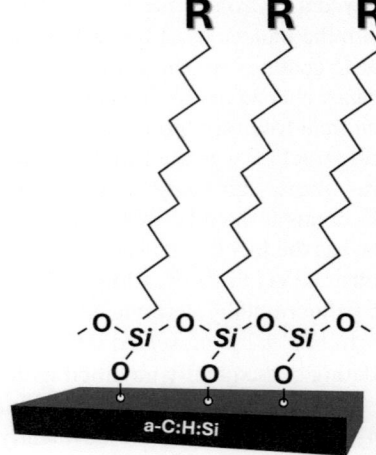

allowed for increased coverage capacity and the production of more crystalline-like layers.

In general the synthesis of self-assembled monolayers from all investigated compounds has been shown to decrease the wear rate and coefficient of friction of Si-DLC, SiO_x-DLC and TiO_x-DLC coatings tested under millinewton load range. A comparative analysis of variations in coefficient of friction for DLC, Si-DLC and Si-DLC modified with DP and PFDP is presented in Fig. 9.5. The relatively high values of the coefficient of friction during the initial testing phase are attributable to the ploughing effect of the counterbody in the organic layers of DP and PFDP.

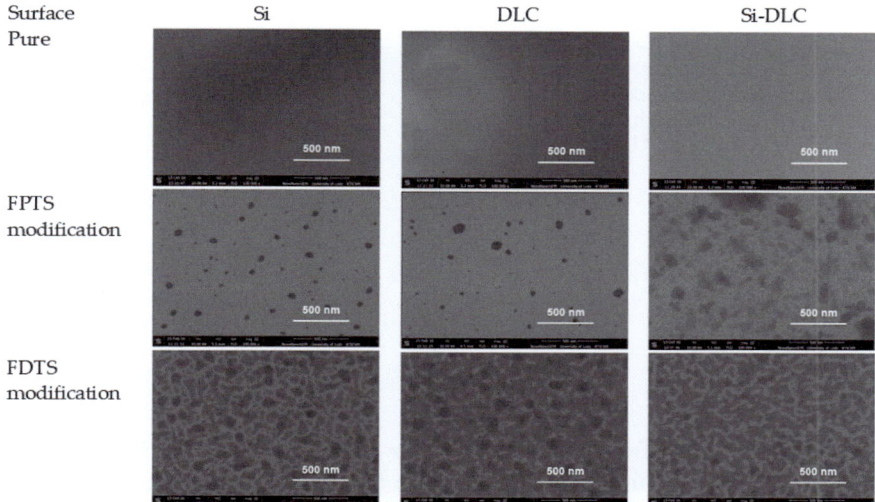

Fig. 9.4 The scanning electron microscopy (SEM) images of silicon, diamond-like carbon layer and silicon incorporated carbon coating after the deposition of FPTS and FDTS SAMs. Reprinted with permission under Creative Commons CC-BY from [14]

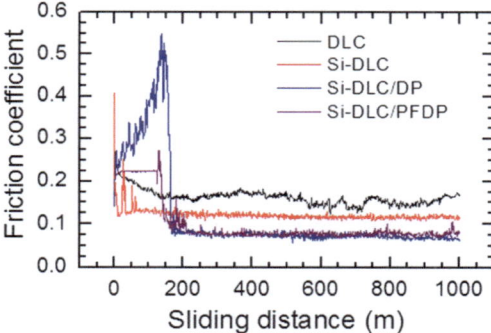

Fig. 9.5 Variation in friction coefficient between DLC, Si-DLC, Si-DLC/DP, Si-DLC/PFDP and Si_3N_4 counterbody (6.35 mm in diameter) under a load of 1N and sliding speed of 0.1 m/s [15]

Subsequent to the removal of the excess of the modifier, the newly emerged thin boundary frictional layers ensured the maintenance of a low and stable CoF.

The processes of friction and wear in a microscale strongly depend on the capillary effects, electrostatic interactions, adhesion forces and work of adhesion [13]. In each case, a substantial decrease in the coefficient of friction was observed following the modification of the original coatings by SAMs. This was associated with the hydrophobicity and low surface free energy of the modified surface. Under low contact pressures, friction forces are predominantly governed by surface interactions, with capillary forces playing a significant role. The presence of adsorbed water molecules at the tribological interface facilitates tribochemical reactions, which can

Fig. 9.6 The effect of FPTS and FDTS modification of Si-DLC (15% Si) coatings on adhesive forces and work of adhesion. Reprinted with permission [13]

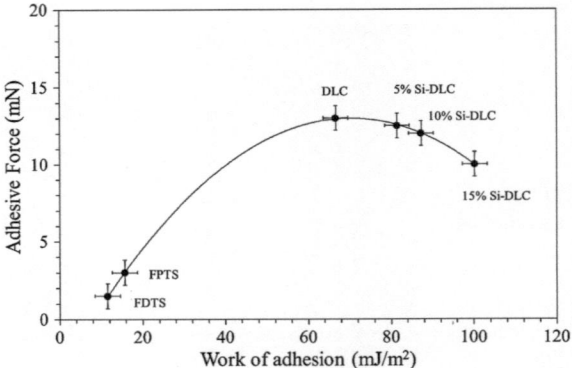

lead to the growth of friction. It has been shown that the presence of hydrophobic self-assembled layers impedes the magnitude of capillary forces and the intensity of tribochemical reactions. Consequently, the values of adhesion force and work of adhesion, i.e. the adhesional components of the friction force, were minimized [14, 15]. The correlation between the adhesive forces and work of adhesion for DLC, Si-DLC and SAMs modified Si-DLC is presented in Fig. 9.6.

The utilisation of organic layers, composed of spontaneously adsorbed molecules on the Si-DLC surface, has emerged as a promising candidate for protective agents. These layers possess a capacity for mitigating friction, adhesion and wear, which are prevalent issues in micro- and nano-electromechanical systems (MEMS/NEMS) employed in the domain of sensors and actuators.

The application of a thin layer of phosphonic acids has been demonstrated to enhance the tribological properties of TiO_x-DLC coatings, as evidenced by a reduction in both the coefficient of friction and the degree of wear [16]. It is noteworthy that the ball-on-disc test results of unmodified TiO_x-DLC coatings exhibited suboptimal performance and durability. A comparison of surface topography and wear track images of TiO_x-DLC, TiO_x-DLC/DP and TiO_x-DLC/PFDP is illustrated in Fig. 9.7.

Ti-DLC coatings modified by DP and PFDP layers exhibited low wear over the entire duration of the test, with an average friction coefficient ranging from 0.17 to 0.19. The obtained wear tracks were smooth after the tests, with only small amounts of wear products gathered on the periphery. Additionally, perfluoro and alkylphosphonic self-assembled monolayers have shown to increase the anticorrosion and antimicrobial properties of TiO_x incorporated DLC coatings [16, 17].

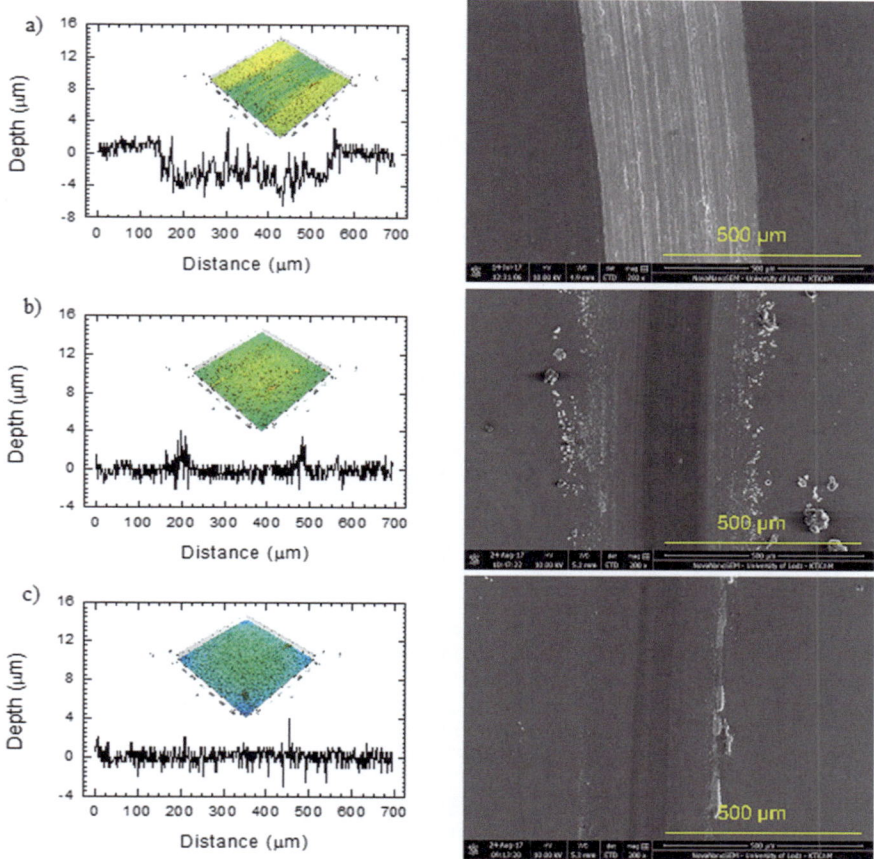

Fig. 9.7 Surface topography and SEM images of wear tracks on Ti-DLC coatings: **a** pure Ti-DLC, **b** DP-modified Ti-DLC and **c** PFDP-modified Ti-DLC. Reprinted with permission from [16]

References

1. Kalin, M., Vizintin, J., Vercammen, K., Barriga, J., Arnšek, A.: The lubrication of DLC coatings with mineral and biodegradable oils having different polar and saturation characteristics. Surf. Coat. Technol. **200**, 4515–4522 (2006). https://doi.org/10.1016/j.surfcoat.2005.03.016
2. Kalin, M., Velkavrh, I., Vižintin, J., Ožbolt, L.: Review of boundary lubrication mechanisms of DLC coatings used in mechanical applications. Meccanica **43**, 623–637 (2008). https://doi.org/10.1007/s11012-008-9149-z
3. Nicosia, C., Huskens, J.: Reactive self-assembled monolayers: from surface functionalization to gradient formation. Mater. Horiz.Horiz. **1**, 32–45 (2014). https://doi.org/10.1039/C3MH00046J
4. Cichomski, M., Kośla, K., Grobelny, J., Kozłowski, W., Szmaja, W.: Tribological and stability investigations of alkylphosphonic acids on alumina surface. Appl. Surf. Sci. **273**, 570–577 (2013). https://doi.org/10.1016/j.apsusc.2013.02.081

5. Cichomski, M.: Tribological investigations of perfluoroalkylsilanes monolayers deposited on titanium surface. Mater. Chem. Phys. **136**, 498–504 (2012). https://doi.org/10.1016/j.matchemphys.2012.07.017
6. Batory, D., Jedrzejczak, A., Kaczorowski, W., Kolodziejczyk, L., Burnat, B.: The effect of Si incorporation on the corrosion resistance of a-C:H:SiOx coatings. Diam. Relat. Mater. **67** (2016). https://doi.org/10.1016/j.diamond.2015.12.002
7. Batory, D., Jedrzejczak, A., Kaczorowski, W., Szymanski, W., Kolodziejczyk, L., Clapa, M., Niedzielski, P.: Influence of the process parameters on the characteristics of silicon-incorporated a-C:H:SiOx coatings. Surf. Coat. Technol. **271**, 112–118 (2015). https://doi.org/10.1016/j.surfcoat.2014.12.073
8. Jedrzejczak, A., Kolodziejczyk, L., Szymanski, W., Piwonski, I., Cichomski, M., Kisielewska, A., Dudek, M., Batory, D.: Friction and wear of a-C:H:SiOx coatings in combination with AISI 316L and ZrO_2 counterbodies. Tribol. Int.. Int. **112**, 155–162 (2017). https://doi.org/10.1016/j.triboint.2017.03.026
9. Jedrzejczak, A., Szymanski, W., Kolodziejczyk, L., Sobczyk-Guzenda, A., Kaczorowski, W., Grabarczyk, J., Niedzielski, P., Kolodziejczyk, A., Batory, D.: Tribological characteristics of a-C:H:Si and a-C:H:SiOx coatings tested in simulated body fluid and protein environment. Materials **15** (2022). https://doi.org/10.3390/ma15062082
10. Batory, D., Jedrzejczak, A., Szymanski, W., Niedzielski, P., Fijalkowski, M., Louda, P., Kotela, I., Hromadka, M., Musil, J.: Mechanical characterization of a-C:H:SiOx coatings synthesized using radio-frequency plasma-assisted chemical vapor deposition method. Thin Solid Films **590** (2015). https://doi.org/10.1016/j.tsf.2015.08.017
11. Jedrzejczak, A., Batory, D., Prowizor, M., Dominik, M., Smietana, M., Cichomski, M., Kisielewska, A., Szymanski, W., Kozlowski, W., Dudek, M.: Titanium (IV) isopropoxide as a source of titanium and oxygen atoms in carbon based coatings deposited by radio frequency plasma enhanced chemical vapour deposition method. Thin Solid Films **693**, 137697 (2020). https://doi.org/10.1016/J.TSF.2019.137697
12. Jedrzejczak, A., Batory, D., Cichomski, M., Miletic, A., Czerniak-Reczulska, M., Niedzielski, P., Dudek, M.: Formation of anatase and srilankite mixture as a result of the thermally induced transformation of the a-C:H:TiOx coating. Surf. Coat. Technol. **400**, 126230 (2020). https://doi.org/10.1016/j.surfcoat.2020.126230
13. Bystrzycka, E., Prowizor, M., Piwoński, I., Kisielewska, A., Batory, D., Jędrzejczak, A., Dudek, M., Kozłowski, W., Cichomski, M.: The effect of fluoroalkylsilanes on tribological properties and wettability of Si-DLC coatings. Mater. Res. Express (2018). https://doi.org/10.1088/2053-1591/aab472
14. Cichomski, M., Borkowska, E., Prowizor, M., Batory, D., Jedrzejczak, A., Dudek, M.: The effect of physicochemical properties of perfluoroalkylsilanes solutions on microtribological features of created self-assembled monolayers. Materials **13**, 3357 (2020). https://doi.org/10.3390/ma13153357
15. Cichomski, M., Kisielewska, A., Prowizor, M., Borkowska, E., Piwoński, I., Dudek, M., Jędrzejczak, A., Batory, D.: The influence of self-assembled monolayers on tribological properties of Si-DLC coatings. Surf. Topogr. **7**, 045006 (2019). https://doi.org/10.1088/2051-672X/ab4347
16. Cichomski, M., Burnat, B., Prowizor, M., Jedrzejczak, A., Batory, D., Piwoński, I., Kozłowski, W., Szymanski, W., Dudek, M.: Tribological and corrosive investigations of perfluoro and alkylphosphonic self-assembled monolayers on Ti incorporated carbon coatings. Tribol. Int.. Int. (2019). https://doi.org/10.1016/j.triboint.2018.10.010
17. Cichomski, M., Prowizor, M., Borkowska, E., Piwoński, I., Jędrzejczak, A., Dudek, M., Batory, D., Wrońska, N., Lisowska, K.: Impact of perfluoro and alkylphosphonic self-assembled monolayers on tribological and antimicrobial properties of Ti-DLC coatings. Materials (2019). https://doi.org/10.3390/ma12152365

MIX
Papier aus verantwortungsvollen Quellen
Paper from responsible sources
FSC® C105338

If you have any concerns about our products,
you can contact us on
ProductSafety@springernature.com

In case Publisher is established outside the EU,
the EU authorized representative is:
**Springer Nature Customer Service Center GmbH
Europaplatz 3, 69115 Heidelberg, Germany**

Printed by Libri Plureos GmbH
in Hamburg, Germany